T0260712

Symmetry Analysis of Differential Equations

Symmetry Analysis of Differential Equations

An Introduction

Daniel J. Arrigo
Department of Mathematics
University of Central Arkansas
Conway, AR

Published by John Wiley & Sons, Inc., Hoboken, New Jersey.
Published simultaneously in Canada.

Library of Congress Cataloging-in-Publication Data:

Arrigo, Daniel J. (Daniel Joseph), 1960-
Symmetry analysis of differential equations : an introduction / Daniel J. Arrigo, Department of Mathematics, University of Central Arkansas, Conway, AR.
pages cm
Includes bibliographical references and index.
ISBN 978-1-118-72140-7 (cloth)
1. Lie groups–Textbooks. 2. Lie groups–Study and teaching (Higher) 3. Lie groups–Study and teaching (Graduate) 4. Differential equations, Partial–Textbooks. I. Title.
QA387.A77 2014
515′.353–dc23

2014007305

10 9 8 7 6 5 4

To the late Bill Ames,
my teacher, my mentor, my friend.

Contents

Preface

Have you ever wondered whether there is some underlying theory that unifies all the seemingly ad-hoc techniques that are used to solve first-order ordinary differential equations (ODEs)? In the 1880s Sophus Lie did [51]. He was able to show that a majority of the techniques used to integrate ODEs could be explained by a theory known as Lie group analysis, where the symmetries of a differential equation could be found and exploited.

This is a self-contained introductory textbook on Lie group analysis, intended for advanced undergraduates and beginning graduate students, or anyone in the science and engineering fields wishing to learn about the use of symmetry methods in solving differential equations. This book has many detailed examples, from the very basic to the more advanced, guiding one through the method of symmetry analysis used for differential equations. The methods presented in this book are very algorithmic in nature, and the author encourages the reader to become familiar with one of the computer algebra packages, such as *Maple*™ or *Mathematica*® to help with the calculations, as they can get extremely long and tedious.

The material presented in this book is based on lectures given by the author over the last 12 years at the University of Central Arkansas (UCA). This book consists of four chapters. In Chapter 1, the reader is introduced to the idea of a symmetry and how these symmetries can leave objects invariant and, in particular, differential equations.

Chapter 2 concentrates on constructing and exploiting symmetries of ordinary differential equations. In particular, the focus is on standard techniques for integrating first-order ODEs: linear, Bernoulli, homogeneous, exact, and Riccati equations are considered. This chapter then considers symmetry methods for second-order equations, higher order equations, and systems of ordinary differential equations.

Chapter 3 extends the ideas to partial differential equations (PDEs). This chapter starts with first-order PDEs and then graduates to second-order PDEs. The power of the method is seen in this chapter, where the heat equation with a source term is considered. This chapter then moves to higher order PDEs and systems of PDEs, where several of the examples (and exercises) have been chosen from various fields of science and engineering.

The last chapter, Chapter 4, starts with a discussion of the nonclassical method–a generalization of Lie's "classical method" and then shows its connection with compatibility. Finally, this chapter ends with a very brief discussion *what's beyond.*

Each chapter has a number of exercises; some are routine while others are more difficult. The latter are denoted by $*$. Many of the answers are given, and for some of the harder or more elaborate problems, a reference to the literature is given.

The material presented in Chapters 1–3 is more than enough for a one-semester course and has been the basis of the course given here at UCA over the last 12 years.

Lastly, for further information and details, including available programs, it is encouraged that you please visit the book's accompanying website at symmetrydes.com.

Acknowledgments

First and foremost, I wish to extend my thanks to W.F. Ames. He was the first to expose me to the wonderful world of the symmetry analysis of differential equations during my PhD studies at Georgia Tech. His patience and support was never ending. I was fortunate to work with Phil Broadbridge and Jim Hill at the University of Wollongong (Australia) for several years while I was a post-doc. I learned so much from them. As this book emerged from lecture notes for a course that I've taught at the University of Central Arkansas, I wish to thank all my students over the past dozen years for their valuable input. In particular, I thank Regan Beckham, David Ekrut, Jackson Fliss, Luis Suazo, and Bode Sule, who were all student researchers and became my coauthors. I would also like to thank David Ekrut and Crystal Spellmann, who read a large portion of this manuscript and gave me valuable suggestions. I would also like to thank Susanne Steitz-Filler with John Wiley & Sons. She helped make this work come to life and helped make the transition so easy.

Finally, I wish to thank my wife Peggy for her patience, undivided support, and encouragement while this project was underway. My love and thanks.

DANIEL J. ARRIGO
University of Central Arkansas
Conway, AR

The lake is calling, I must go.

CHAPTER 1

An Introduction

1.1 WHAT IS A SYMMETRY?

What is a symmetry? A symmetry is a transformation that leaves an object unchanged or "invariant." For example, if we start with a basic equilateral triangle with the vertices labeled as 1, 2, and 3 (Figure 1.1), then a reflection through any one of three different bisection axes (Figure 1.2) or rotations through angles of $\frac{2\pi}{3}$ and $\frac{4\pi}{3}$ (Figure 1.3) leaves the triangle invariant.

Another example is the rotation of a disk through an angle ε. Consider the points (x, y) and $(\bar{x}, \bar{y})$, on the circumference of a circle of radius r (Figure 1.4). We can write these in terms of the radius and the angles θ (a reference angle) and $\theta + \varepsilon$, (after rotation), that is,

These then become

$$x = r\cos\theta, \quad \bar{x} = r\cos(\theta + \varepsilon), \tag{1.1a}$$

$$y = r\sin\theta, \quad \bar{y} = r\sin(\theta + \varepsilon), \tag{1.1b}$$

or, after eliminating θ

$$\bar{x} = x\cos\varepsilon - y\sin\varepsilon \tag{1.2a}$$

$$\bar{y} = y\cos\varepsilon + x\sin\varepsilon. \tag{1.2b}$$

Symmetry Analysis of Differential Equations: An Introduction,
First Edition. Daniel J. Arrigo.

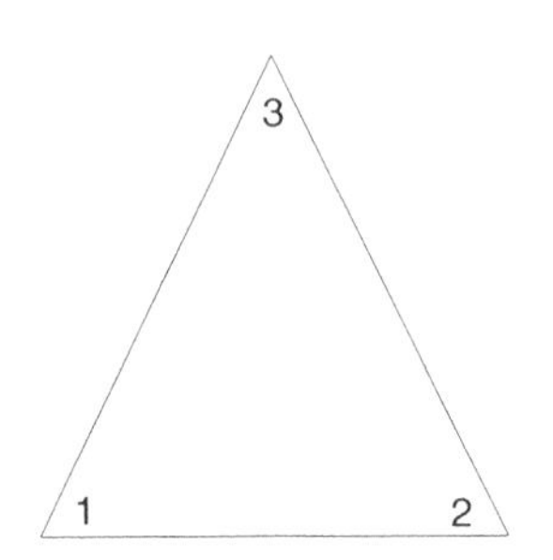

FIGURE 1.1 An equilateral triangle.

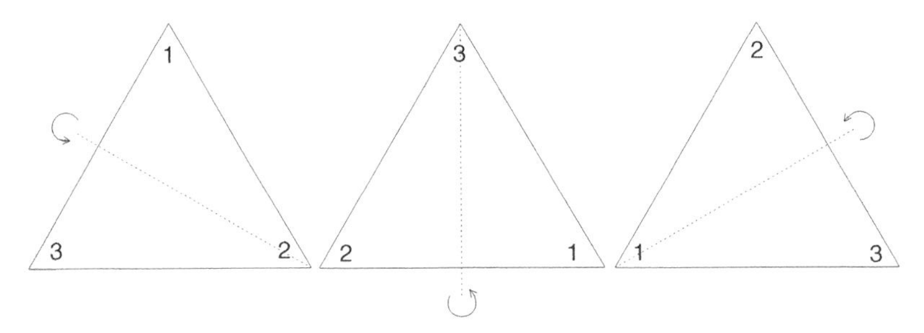

FIGURE 1.2 Reflections of an equilateral triangle.

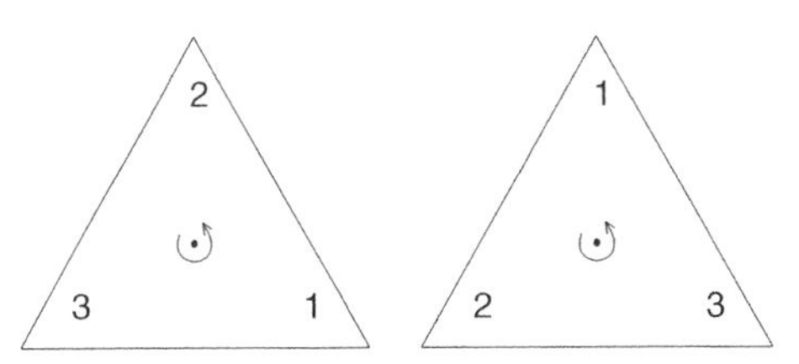

FIGURE 1.3 Rotations of an equilateral triangle through $\frac{2\pi}{3}$ and $\frac{4\pi}{3}$.

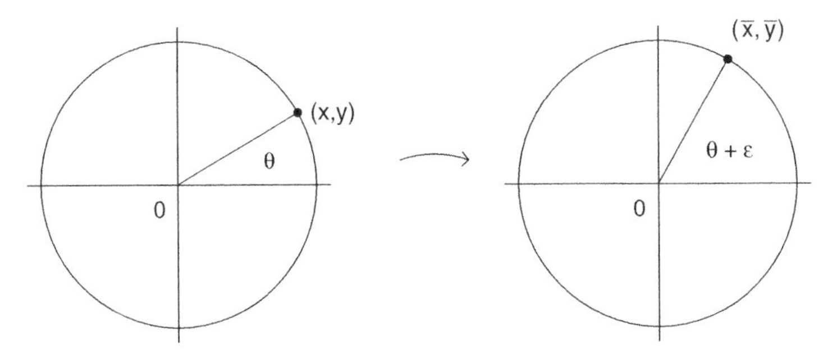

FIGURE 1.4 Rotation of a circle.

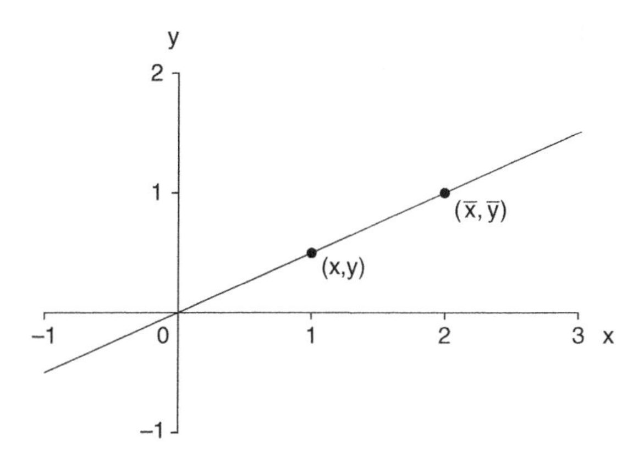

FIGURE 1.5 Invariance of the line $y = \frac{1}{2}x$.

To show invariance of the circle under (1.2) is to show that $\overline{x}^2 + \overline{y}^2 = r^2$ if $x^2 + y^2 = r^2$. Therefore,

$$\begin{aligned}\overline{x}^2 + \overline{y}^2 &= (x\cos\varepsilon - y\sin\varepsilon)^2 + (y\cos\varepsilon + x\sin\varepsilon)^2 \\ &= x^2\cos^2\varepsilon - 2xy\sin\varepsilon\cos\varepsilon + y^2\sin^2\varepsilon \\ &\quad + x^2\sin^2\varepsilon + 2xy\sin\varepsilon\cos\varepsilon + y^2\cos^2\varepsilon \\ &= x^2 + y^2 \\ &= r^2.\end{aligned}$$

As a third example, consider the line $y = \frac{1}{2}x$ and the transformation

$$\overline{x} = e^{\varepsilon}x \quad \overline{y} = e^{\varepsilon}y. \tag{1.3}$$

The line is invariant under (1.3) as (Figure 1.5)

$$\overline{y} = \frac{1}{2}\overline{x} \text{ then } e^{\varepsilon}y = \frac{1}{2}e^{\varepsilon}x \text{ if } y = \frac{1}{2}x.$$

EXAMPLE 1.1

Show the equation

$$x^2y^2 - xy^2 + 2xy - y^2 - y + 1 = 0 \tag{1.4}$$

is invariant under

$$\bar{x} = x + \varepsilon, \quad \bar{y} = \frac{y}{1 - \varepsilon y}. \tag{1.5}$$

For this example, it is actually easier to rewrite (1.4) as

$$\left(x + \frac{1}{y}\right)^2 - \left(x + \frac{1}{y}\right) - 1 = 0. \tag{1.6}$$

Under the transformation (1.5), the term $x + \dfrac{1}{y}$ becomes

$$\bar{x} + \frac{1}{\bar{y}} = x + \varepsilon + \frac{1 - \varepsilon y}{y} = x + \varepsilon + \frac{1}{y} - \varepsilon = x + \frac{1}{y} \tag{1.7}$$

and invariance of (1.6) readily follows.

It is important to realize that not all equations are invariant under all transformations. Consider $y - 1 = 3(x - 1)$ and the transformation (1.3) again. If this were invariant, then

$$\bar{y} - 1 = 3\left(\bar{x} - 1\right) \quad \text{if} \quad y - 1 = 3(x - 1).$$

Upon substitution, we have $e^{\varepsilon} y - 1 = 3(e^{\varepsilon} x - 1)$, which is clearly not the original line and hence not invariant under (1.3).

The transformations (1.2), (1.3), and (1.5) are very special and are referred to as *Lie transformation groups* or just *Lie groups*. ■

1.2 LIE GROUPS

In general, we consider transformations

$$\bar{x}_i = f_i(x_j, \varepsilon), \quad i, j = 1, 2, 3, \cdots n.$$

These are called a *one-parameter Lie group*, where ε is the parameter. First and foremost, they form a group. That is, they satisfy the following axioms, where G is the group and $\phi(a, b)$ the law of composition.

1. *Closure*. If $a, b \in G$, then $\phi(a, b) \in G$.
2. *Associative*. If $a, b, c \in G$, then $\phi(a, \phi(b, c)) = \phi(\phi(a, b), c)$.
3. *Identity*. If $a \in G$, then there exists an $e \in G$ such that $\phi(a, e) = \phi(e, a) = a$.

4. *Inverse*. If $a \in G$, then there exists a unique element $a^{-1} \in G$ such that $\phi(a, a^{-1}) = \phi(a^{-1}, a) = e$.

Second, they further satisfy the following properties:

1. f_i is a smooth function of the variables x_j.
2. f_i is analytic function in the parameter ε, that is, a function with a convergent Taylor series in ε.
3. $\varepsilon = 0$ can always be chosen to correspond with the identity element e.
4. the law of composition can be taken as $\phi(a, b) = a + b$.

Our focus is on transformation groups, so our discussion is confined to these types of groups.

EXAMPLE 1.2

Consider (Figure 1.6)

$$\bar{x} = ax, \quad a \in \mathbb{R}\backslash\{0\}. \tag{1.8}$$

1. *Closure*. If $\bar{x} = ax$ and $\tilde{x} = b\bar{x}$, then $\tilde{x} = abx$. In this example, the law of composition is $\phi(a, b) = ab$.
2. *Associative*. As $\phi(a, b) = ab$, then $\phi(a, \phi(b, c)) = a(bc) = (ab)c = \phi(\phi(a, b), c)$.
3. *Identity*. In this case, $e = 1$ as $\phi(a, 1) = a1 = a$.
4. *Inverse*. Here $a^{-1} = \dfrac{1}{a}$ as $\phi(a, \frac{1}{a}) = 1$.

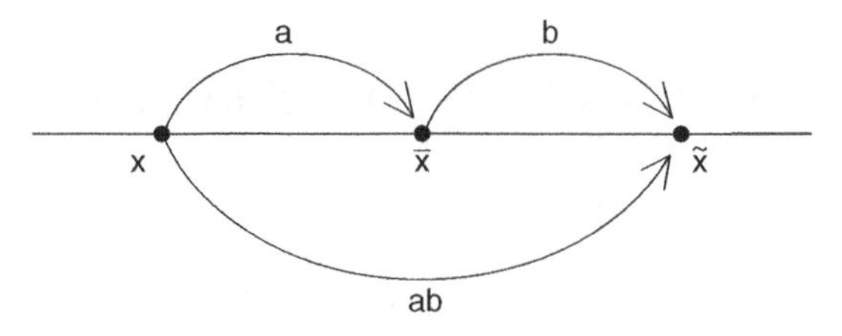

FIGURE 1.6 Scaling group $\bar{x} = ax$ and its composition.

We note that if we reparameterize the group by letting $a = e^{\varepsilon}$, then the group becomes a Lie group. ■

EXAMPLE 1.3

Consider

$$\overline{x} = \frac{xy}{y-\varepsilon}, \quad \overline{y} = y - \varepsilon, \quad \varepsilon \in \mathbb{R}. \tag{1.9}$$

1. *Closure*. If
$$\overline{x} = \frac{xy}{y-a} \quad \text{and} \quad \overline{y} = y - a$$
and if
$$\tilde{x} = \frac{\overline{xy}}{\overline{y}-b} \quad \text{and} \quad \tilde{y} = \overline{y} - b,$$
then
$$\tilde{x} = \frac{\dfrac{xy}{y-a}\cdot(y-a)}{y-a-b} = \frac{xy}{y-(a+b)}$$
and
$$\tilde{y} = y - (a+b).$$
In this example, the law of composition is $\phi(a,b) = a + b$.
2. *Associative*. As $\phi(a,b) = a + b$, then $\phi(a,\phi(b,c)) = a + (b + c) = (a + b) + c = \phi(\phi(a,b),c)$.
3. *Identity*. In this case, $e = 0$.
4. *Inverse*. Here $a^{-1} = -\varepsilon$ as $\phi(\varepsilon, -\varepsilon) = 0$.

This is an example of a Lie group. It is an easy matter to show that $xy = 1$ is invariant under (1.9). Figure 1.7 illustrates the composition of two successive transformations ■

1.3 INVARIANCE OF DIFFERENTIAL EQUATIONS

We are starting to discover that equations can be invariant under a Lie group. This leads us to the following question: Can differential equations be invariant under Lie groups? The following examples illustrate an answer to that question. Consider the simple differential equation

$$\frac{dy}{dx} = xy^3, \tag{1.10}$$

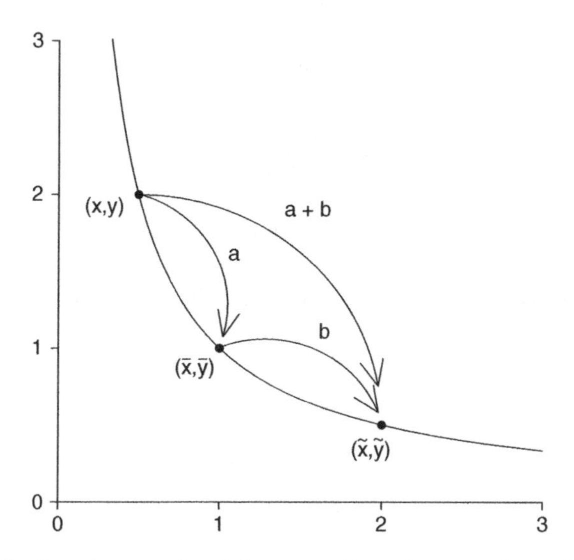

FIGURE 1.7 Equation $xy = 1$ and the composition of two transformations of (1.9).

and the Lie group

$$\overline{x} = \mathrm{e}^{\varepsilon} x, \quad \overline{y} = \mathrm{e}^{-\varepsilon} y. \tag{1.11}$$

Is the ODE (1.10) invariant under (1.11)? It is an easy matter to calculate

$$\frac{\mathrm{d}\overline{y}}{\mathrm{d}\overline{x}} = \mathrm{e}^{-2\varepsilon} \frac{\mathrm{d}y}{\mathrm{d}x} \quad \text{and} \quad \overline{x}\overline{y}^3 = \mathrm{e}^{-2\varepsilon} xy^3 \tag{1.12a}$$

and clearly, under (1.11), (1.10) is invariant, as from (1.12) we see that

$$\frac{\mathrm{d}\overline{y}}{\mathrm{d}\overline{x}} = \overline{x}\,\overline{y}^3 \quad \text{since} \quad \frac{\mathrm{d}y}{\mathrm{d}x} = xy^3.$$

EXAMPLE 1.4

Show

$$\frac{\mathrm{d}y}{\mathrm{d}x} = \frac{(xy+1)^3}{x^5} + \frac{1}{x^2} \tag{1.13}$$

is invariant under

$$\overline{x} = \frac{x}{1+\varepsilon x}, \quad \overline{y} = y - \varepsilon. \tag{1.14}$$

We first calculate $\frac{d\bar{y}}{d\bar{x}}$ by the chain rule

$$\begin{aligned}\frac{d\bar{y}}{d\bar{x}} &= \frac{d\bar{y}}{dx}\Big/\frac{d\bar{x}}{dx} \\ &= \frac{dy}{dx}(1+\varepsilon x)^2. \end{aligned} \tag{1.15}$$

Next, we focus on the first term on the right-hand side of (1.13). So

$$\begin{aligned}\frac{(\bar{x}\bar{y}+1)^3}{\bar{x}^5} &= \frac{\left(\frac{x}{1+\varepsilon x}\cdot(y-\varepsilon)+1\right)^3}{\left(\frac{x}{1+\varepsilon x}\right)^5} \\ &= \frac{\left(\frac{xy+1}{1+\varepsilon x}\right)^3}{\left(\frac{x}{1+\varepsilon x}\right)^5} \\ &= \frac{(xy+1)^3}{x^5}(1+\varepsilon x)^2. \end{aligned}$$

Thus, the entire right-hand side of (1.13) becomes

$$\frac{(\bar{x}\bar{y}+1)^3}{\bar{x}^5} + \frac{1}{\bar{x}^2} = \frac{(xy+1)^3}{x^5}(1+\varepsilon x)^2 + \frac{(1+\varepsilon x)^2}{x^2}. \tag{1.16}$$

To show invariance is to show that

$$\frac{d\bar{y}}{d\bar{x}} = \frac{(\bar{x}\bar{y}+1)^3}{\bar{x}^5} + \frac{1}{\bar{x}^2} \quad \text{if} \quad \frac{dy}{dx} = \frac{(xy+1)^3}{x^5} + \frac{1}{x^2}. \tag{1.17}$$

Using (1.15) and (1.16) in (1.17) shows (1.17) to be true.

We will now turn our attention to solving some differential equations. ■

1.4 SOME ORDINARY DIFFERENTIAL EQUATIONS

Consider the Riccati equation

$$\frac{dy}{dx} = y^2 - \frac{y}{x} - \frac{1}{x^2}. \tag{1.18}$$

Typically, to solve this ODE, we need one solution. It is an easy matter to show that if we guess a solution of the form $y = \frac{k}{x}$, then

$$k = \pm 1.$$

If we choose

$$y_1 = \frac{1}{x}$$

and let

$$y = \frac{1}{u} + \frac{1}{x}, \tag{1.19}$$

where $u = u(x)$, then substituting into equation (1.18) and simplifying gives

$$u' + \frac{u}{x} = -1,$$

which is linear. The integrating factor μ is

$$\mu = \mathrm{e}^{\int \frac{\mathrm{d}x}{x}} = x,$$

and so the linear equation is easily integrated, giving

$$u = \frac{c - x^2}{2x},$$

where c is an arbitrary constant of integration. Substituting this into (1.19) gives

$$y = \frac{2x}{c - x^2} + \frac{1}{x}. \tag{1.20}$$

We find that the procedure is long and we do need one solution to find the general solution of (1.18). However, if we let

$$x = \mathrm{e}^s, \quad y = r\mathrm{e}^{-s}, \tag{1.21}$$

where $s = s(r)$, then substituting into (1.18) gives

$$\frac{\mathrm{e}^{-s} - r\mathrm{e}^{-s}s'}{\mathrm{e}^s s'} = r^2\mathrm{e}^{-2s} - r\mathrm{e}^{-2s} - \mathrm{e}^{-2s}$$

and solving for s' gives

$$\frac{ds}{dr} = \frac{1}{r^2 - 1}, \tag{1.22}$$

an equation which is separable and independent of s! This is easily integrated and using (1.21) gives rise to the solution (1.20).

Consider

$$\frac{dy}{dx} = \frac{y}{x} + \frac{x^2}{x + y}. \tag{1.23}$$

Unfortunately, there is no simple way to solve this ODE. However, if we let

$$x = s, \quad y = rs, \tag{1.24}$$

where again $s = s(r)$, then (1.23) becomes

$$\frac{ds}{dr} = r + 1, \tag{1.25}$$

an equation which is also separable and independent of s! Again, it is easily integrated, giving

$$s = \frac{1}{2}r^2 + r + c, \tag{1.26}$$

and using (1.24) gives rise to the solution of (1.23)

$$x = \frac{1}{2}\frac{y^2}{x^2} + \frac{y}{x} + c. \tag{1.27}$$

Finally, we consider

$$\frac{dy}{dx} = \frac{2y^3(x - y - xy)}{x(x - y)^2}. \tag{1.28}$$

This is a complicated ODE without a standard way of solving it. However, under the change of variables

$$x = \frac{1}{r + s}, \quad y = \frac{1}{s}, \tag{1.29}$$

(1.28) becomes

$$\frac{ds}{dr} = -\frac{2(r + 1)}{r^2 + 2r + 2}, \tag{1.30}$$

again, an equation which is separable and independent of s! Integrating (1.30) gives

$$s = -\ln|r^2 + 2r + 2| + c \tag{1.31}$$

and via (1.29) gives

$$\frac{1}{y} = -\ln\left|\left(\frac{1}{x} - \frac{1}{y}\right)^2 + 2\left(\frac{1}{x} - \frac{1}{y}\right) + 2\right| + c, \tag{1.32}$$

the exact solution of (1.28).

In summary, we have considered three different ODEs and have shown that by introducing new variables, these ODEs can be reduced to new ODEs (Table 1.1) that are separable and independent of s.

We are naturally led to the following questions:

1. What do these three seemingly different ODEs have in common?
2. How did I know to pick the new coordinates (r, s) (if they even exist) so that the original equation reduces to one that is separable and independent of s?

The answer to the first question is that all of the ODEs are invariant under some Lie group. The first ODE

$$\frac{dy}{dx} = y^2 - \frac{y}{x} - \frac{1}{x^2} \tag{1.33}$$

TABLE 1.1
Equation (1.18), (1.23) and (1.28) and their separability

Equation	Transformation	New Equation
$\frac{dy}{dx} = y^2 - \frac{y}{x} - \frac{1}{x^2}$	$x = e^s, \quad y = re^{-s}$	$\frac{ds}{dr} = \frac{1}{r^2 - 1}$
$\frac{dy}{dx} = \frac{y}{x} + \frac{x^2}{x + y}$	$x = s, \quad y = rs$	$\frac{ds}{dr} = r + 1$
$\frac{dy}{dx} = \frac{2y^3(x - y - xy)}{x(x - y)^2}$	$x = \frac{1}{r + s}, \quad y = \frac{1}{s}$	$\frac{ds}{dr} = -\frac{2(r + 1)}{r^2 + 2r + 2}$

is invariant under

$$\bar{x} = e^{\varepsilon} x, \quad \bar{y} = e^{-\varepsilon} y, \tag{1.34}$$

the second ODE

$$\frac{dy}{dx} = \frac{y}{x} + \frac{x^2}{x+y} \tag{1.35}$$

is invariant under

$$\bar{x} = x + \varepsilon, \quad \bar{y} = \frac{(x+\varepsilon)\,y}{x}, \tag{1.36}$$

and the third ODE

$$\frac{dy}{dx} = \frac{2y^3(x - y - xy)}{x(x-y)^2} \tag{1.37}$$

is invariant under

$$\bar{x} = \frac{x}{1+\varepsilon x}, \quad \bar{y} = \frac{y}{1+\varepsilon y}. \tag{1.38}$$

The answer to the second question will be revealed in Chapter 2!

EXERCISES

1. Show that the following are Lie groups:

$$\begin{aligned}
&(i) \quad \bar{x} = e^{\varepsilon} x,\\
&(ii) \quad \bar{x} = \sqrt{x^2 + \varepsilon},\\
&(iii) \quad \bar{x} = \frac{(y+\varepsilon)x}{y}, \quad \bar{y} = y + \varepsilon,\\
&(iv) \quad \bar{x} = \frac{x}{1+\varepsilon x}, \quad \bar{y} = \frac{y}{1+\varepsilon y}.
\end{aligned}$$

2. Show that the following equations are invariant under the given Lie group

$$\begin{aligned}
&(i) \quad x^2y^2 + e^{xy} = 1 + xy, && \bar{x} = e^{\varepsilon} x, \quad \bar{y} = e^{-\varepsilon} y,\\
&(ii) \quad y^4 + 2xy^2 + x^2 + 2y^2 + 2x = 0, && \bar{x} = x - \varepsilon, \quad \bar{y} = \sqrt{y^2 + \varepsilon},\\
&(iii) \quad x^2 - y^2 - 2xy \sin\frac{y}{x} = 0, && \bar{x} = \frac{x}{1+\varepsilon x}, \quad \bar{y} = \frac{y}{1+\varepsilon x}.
\end{aligned}$$

3. Find functions $a(\varepsilon)$, $b(\varepsilon)$, $c(\varepsilon)$, and $d(\varepsilon)$ such that

$$y - y_0 = m(x - x_0) \tag{1.39}$$

is invariant under

$$\bar{x} = a(\varepsilon)x + b(\varepsilon), \quad \bar{y} = c(\varepsilon)y + d(\varepsilon). \tag{1.40}$$

Does this form a Lie group? If not, find the form of (1.40) that not only leaves (1.39) invariant but also forms a Lie group.

4. Show that the following ODEs are invariant under the given Lie groups

$$(i) \quad \frac{\mathrm{d}y}{\mathrm{d}x} = 2y^2 + xy^3, \qquad \bar{x} = \mathrm{e}^{\varepsilon}x, \quad \bar{y} = \mathrm{e}^{-\varepsilon}y,$$

$$(ii) \quad \frac{\mathrm{d}y}{\mathrm{d}x} = \frac{x^2 y}{x^3 + xy + y^2}, \qquad \bar{x} = \frac{x}{1+\varepsilon y}, \quad \bar{y} = \frac{y}{1+\varepsilon y},$$

$$(iii) \quad \frac{\mathrm{d}y}{\mathrm{d}x} = \frac{y^2}{x^2}F\left(\frac{1}{x} - \frac{1}{y}\right), \quad \bar{x} = \frac{x}{1+\varepsilon x}, \quad \bar{y} = \frac{y}{1+\varepsilon y}.$$

5. Find the constants a and b such that the ODE

$$\frac{\mathrm{d}y}{\mathrm{d}x} = \frac{3xy + 2y^3}{x^2 + 3xy^2},$$

is invariant under the Lie group of transformations

$$\bar{x} = \mathrm{e}^{a\varepsilon}x, \quad \bar{y} = \mathrm{e}^{b\varepsilon}y.$$

6. (i) Show that

$$\frac{\mathrm{d}y}{\mathrm{d}x} = \frac{1 - y - 2xy^2}{x(2xy+1)}$$

is invariant under

$$\bar{x} = x + \varepsilon, \quad \bar{y} = \frac{xy}{x+\varepsilon}.$$

(ii) Show that under the change of variables

$$r = xy, \quad s = x,$$

the original equation becomes

$$\frac{\mathrm{d}s}{\mathrm{d}r} = 2r + 1.$$

7. (i) Show that

$$\frac{\mathrm{d}y}{\mathrm{d}x} = \frac{y^2 + (x - 2x^2)y - x^3}{x(x + y)}$$

is invariant under

$$\overline{x} = x + \varepsilon, \quad \overline{y} = \frac{(x + \varepsilon)y}{x}.$$

(ii) Show that under the change of variables

$$r = \frac{y}{x}, \qquad s = x,$$

the original equation becomes

$$\frac{\mathrm{d}s}{\mathrm{d}r} = -\frac{r + 1}{2r + 1}.$$

8. (i) Show that

$$\frac{\mathrm{d}y}{\mathrm{d}x} = F(x) \tag{1.41}$$

is invariant under

$$\overline{x} = x, \quad \overline{y} = y + \varepsilon. \tag{1.42}$$

(ii) Prove that the only ordinary differential equation of the form

$$\frac{\mathrm{d}y}{\mathrm{d}x} = F(x, y)$$

that is invariant under (1.42) is of the form of (1.41).

CHAPTER 2

Ordinary Differential Equations

In this chapter, we focus on the symmetries of ordinary differential equations. At the close of Chapter 1, we saw three different ODEs, (1.18), (1.23), and (1.28), which were all transformed to equations that were separable and independent of s. We also stated that each was invariant under some Lie group (see Table 2.1).

We now ask, where did these transformations come from and how do they relate to Lie groups? In an attempt to answer this question, we consider these transformations in detail.

EXAMPLE 2.1

The ODE

$$\frac{dy}{dx} = y^2 - \frac{y}{x} - \frac{1}{x^2} \tag{2.1}$$

is invariant under

$$\bar{x} = e^{\varepsilon} x, \quad \bar{y} = e^{-\varepsilon} y, \tag{2.2}$$

as

$$\frac{d\bar{y}}{d\bar{x}} = \bar{y}^2 - \frac{\bar{y}}{\bar{x}} - \frac{1}{\bar{x}^2}$$

Symmetry Analysis of Differential Equations: An Introduction,
First Edition. Daniel J. Arrigo.

TABLE 2.1
Invariance of equations (1.18), (1.23) and (1.28)

Equation	Transformation	Lie Group
$\dfrac{dy}{dx} = y^2 - \dfrac{y}{x} - \dfrac{1}{x^2}$	$x = e^s, \quad y = re^{-s}$	$\bar{x} = e^{\varepsilon} x, \quad \bar{y} = e^{-\varepsilon} y$
$\dfrac{dy}{dx} = \dfrac{y}{x} + \dfrac{x^2}{x+y}$	$x = s, \quad y = rs$	$\bar{x} = x + \varepsilon, \quad \bar{y} = \dfrac{(x+\varepsilon)y}{x}$
$\dfrac{dy}{dx} = \dfrac{2y^3(x - y - xy)}{x(x-y)^2}$	$x = \dfrac{1}{r+s}, \quad y = \dfrac{1}{s}$	$\bar{x} = \dfrac{x}{1+\varepsilon x}, \quad \bar{y} = \dfrac{y}{1+\varepsilon y}$

$$
\begin{aligned}
e^{-2\varepsilon} \frac{dy}{dx} &= e^{-2\varepsilon} \left(y^2 - \frac{y}{x} - \frac{1}{x^2} \right) \\
\frac{dy}{dx} &= y^2 - \frac{y}{x} - \frac{1}{x^2}.
\end{aligned}
$$

Further, under the change of variables

$$x = e^s, \quad y = re^{-s} \tag{2.3}$$

then (2.1) becomes

$$\frac{ds}{dr} = \frac{1}{r^2 + 1}, \tag{2.4}$$

noting that (2.4), as with all equations of the form $\dfrac{ds}{dr} = G(r)$ is invariant under

$$\bar{r} = r, \quad \bar{s} = s + \varepsilon, \tag{2.5}$$

(cf. Exercise Chapter 1, #8). It is interesting to note that the transformation (2.3) is invariant under the combined use of (2.2) and (2.5). To see this

$$
\begin{aligned}
\bar{x} &= e^{\bar{s}} & \bar{y} &= \bar{r} e^{-\bar{s}} \\
e^{\varepsilon} x &= e^{s+\varepsilon} & e^{-\varepsilon} y &= r e^{-(s+\varepsilon)} \\
e^{\varepsilon} x &= e^{s} e^{\varepsilon} & e^{-\varepsilon} y &= r e^{-s} e^{-\varepsilon} \\
x &= e^{s} & y &= r e^{-s}.
\end{aligned}
$$

■

EXAMPLE 2.2

From the previous chapter, we saw that

$$\frac{dy}{dx} = \frac{y}{x} + \frac{x^2}{x+y} \tag{2.6}$$

was invariant under

$$\bar{x} = x + \varepsilon, \quad \bar{y} = \frac{(x+\varepsilon)\,y}{x}, \tag{2.7}$$

and under the change of variables

$$x = s, \quad y = rs, \tag{2.8}$$

(2.6) was transformed into

$$\frac{\mathrm{d}s}{\mathrm{d}r} = r + 1.$$

It is an easy matter to verify that (2.8) is invariant under (2.7) and (2.5) as

$$\begin{aligned} \bar{x} &= \bar{s} & \bar{y} &= \bar{r}\,\bar{s} \\ x + \varepsilon &= s + \varepsilon & \frac{(x+\varepsilon)\,y}{x} &= r\,(s+\varepsilon) \\ x &= s & y &= r\,s, \end{aligned} \tag{2.9}$$

noting that we used $x = s$ in the second column of (2.9). ■

EXAMPLE 2.3

Finally, we recall from the previous chapter that

$$\frac{\mathrm{d}y}{\mathrm{d}x} = \frac{2y^3(x - y - xy)}{x(x-y)^2} \tag{2.10}$$

was invariant under the Lie group

$$\bar{x} = \frac{x}{1+\varepsilon x}, \quad \bar{y} = \frac{y}{1+\varepsilon y}, \tag{2.11}$$

and under the change of variables

$$x = \frac{1}{r+s}, \quad y = \frac{1}{s} \tag{2.12}$$

(2.10) was transformed to

$$\frac{\mathrm{d}s}{\mathrm{d}r} = -\frac{2r+2}{r^2+2r+1}.$$

We leave it to the reader to show that the transformation (2.12) itself is also invariant under (2.11) and (2.5).

This suggests that if we had a Lie group that left a particular differential equation invariant, say

$$\bar{x} = f(x, y, \varepsilon), \quad \bar{y} = g(x, y, \varepsilon), \tag{2.13}$$

then the change of variables

$$x = A(r, s), \quad y = B(r, s), \tag{2.14}$$

would lead to a separable equation that would be invariant under (2.13) and (2.5), that is,

$$\bar{x} = A(\bar{r}, \bar{s}), \quad \bar{y} = B(\bar{r}, \bar{s}) \tag{2.15}$$

if (2.14) holds. Combining (2.5), (2.13), and (2.15) gives

$$f(x, y, \varepsilon) = A(r, s+\varepsilon), \quad g(x, y, \varepsilon) = B(r, s+\varepsilon) \tag{2.16}$$

and further, using (2.14) gives (2.16) as

$$f(A(r,s), B(r,s), \varepsilon) = A(r, s+\varepsilon), \tag{2.17a}$$

$$g(A(r,s), B(r,s), \varepsilon) = B(r, s+\varepsilon), \tag{2.17b}$$

two functional equations for A and B. However, solving (2.17) really hinges on the fact that we have the Lie group that leaves our differential invariant. If we require our differential equation

$$\frac{dy}{dx} = F(x, y)$$

to be invariant under (2.13) then

$$\frac{d\bar{y}}{d\bar{x}} = F(\bar{x}, \bar{y}) \quad \Rightarrow \quad \frac{g_x + g_y F(x,y)}{f_x + f_y F(x,y)} = F(f(x,y,\varepsilon), g(x,y,\varepsilon)),$$

a nonlinear PDE for f and g that is really too hard to solve! So at this point it seems rather hopeless. However, Lie considered not the Lie group itself but an expanded form of the Lie group which are called *infinitesimal transformations*. ■

2.1 INFINITESIMAL TRANSFORMATIONS

Consider

$$\bar{x} = f(x, y, \varepsilon), \quad \bar{y} = g(x, y, \varepsilon), \tag{2.18}$$

with

$$f(x, y, 0) = x \text{ and } g(x, y, 0) = y. \tag{2.19}$$

If we assume that ε is small, then we construct a Taylor series of (2.18) about $\varepsilon = 0$. Thus,

$$\bar{x} = f(x, y, 0) + \left.\frac{\partial f}{\partial \varepsilon}\right|_{\varepsilon=0} \varepsilon + O(\varepsilon^2), \tag{2.20a}$$

$$\bar{y} = g(x, y, 0) + \left.\frac{\partial g}{\partial \varepsilon}\right|_{\varepsilon=0} \varepsilon + O(\varepsilon^2). \tag{2.20b}$$

If we let

$$\left.\frac{\partial f}{\partial \varepsilon}\right|_{\varepsilon=0} = X(x, y), \quad \left.\frac{\partial g}{\partial \varepsilon}\right|_{\varepsilon=0} = Y(x, y) \tag{2.21}$$

and use (2.19), then (2.20) becomes

$$\bar{x} = x + X(x, y)\varepsilon + O(\varepsilon^2), \quad \bar{y} = y + Y(x, y)\varepsilon + O(\varepsilon^2). \tag{2.22}$$

These are referred to as *infinitesimal transformations* and X and Y as simply *infinitesimals*.

Here we consider two examples where the infinitesimals are obtained from the Lie group.

EXAMPLE 2.4

Consider the Lie group

$$\bar{x} = \mathrm{e}^{\varepsilon} x, \quad \bar{y} = \mathrm{e}^{-\varepsilon} y.$$

Therefore,

$$\frac{\partial \bar{x}}{\partial \varepsilon} = \mathrm{e}^{\varepsilon} x \quad \text{so} \quad \left.\frac{\partial \bar{x}}{\partial \varepsilon}\right|_{\varepsilon=0} = x,$$

$$\frac{\partial \bar{y}}{\partial \varepsilon} = -\mathrm{e}^{-\varepsilon} y \quad \text{so} \quad \left.\frac{\partial \bar{y}}{\partial \varepsilon}\right|_{\varepsilon=0} = -y,$$

and

$$X(x, y) = x, \quad Y(x, y) = -y. \tag{2.23}$$

■

EXAMPLE 2.5

Consider the Lie group

$$\bar{x} = x\cos\varepsilon - y\sin\varepsilon, \quad \bar{y} = y\cos\varepsilon + x\sin\varepsilon.$$

Therefore,

$$\frac{\partial\bar{x}}{\partial\varepsilon} = -x\sin\varepsilon - y\cos\varepsilon \quad \text{so} \quad \left.\frac{\partial\bar{x}}{\partial\varepsilon}\right|_{\varepsilon=0} = -y,$$

$$\frac{\partial\bar{y}}{\partial\varepsilon} = -y\sin\varepsilon + x\cos\varepsilon \quad \text{so} \quad \left.\frac{\partial\bar{y}}{\partial\varepsilon}\right|_{\varepsilon=0} = x.$$

Thus,

$$X(x, y) = -y, \quad Y(x, y) = x.$$

If we had the infinitesimals X and Y, we could recover the Lie group from which they came. We would solve the system of differential equations

$$\frac{d\bar{x}}{d\varepsilon} = X(\bar{x}, \bar{y}), \quad \bar{x}|_{\varepsilon=0} = x, \tag{2.24a}$$

$$\frac{d\bar{y}}{d\varepsilon} = Y(\bar{x}, \bar{y}), \quad \bar{y}|_{\varepsilon=0} = y, \tag{2.24b}$$

(see Bluman and Kumei [1] for the proof). The following examples illustrate this.

■

EXAMPLE 2.6

Consider $X = 1$ and $Y = 2x$. From (2.24) we are required to solve

$$\frac{d\bar{x}}{d\varepsilon} = 1, \quad \frac{d\bar{y}}{d\varepsilon} = 2\bar{x}, \tag{2.25}$$

subject to the initial conditions given in (2.24). The solution of the first is

$$\bar{x} = \varepsilon + c(x, y),$$

where c is an arbitrary function of its arguments. If we impose the initial condition $\bar{x} = x$ when $\varepsilon = 0$ gives $c(x, y) = x$, leads to

$$\bar{x} = x + \varepsilon. \tag{2.26}$$

Using (2.26) in the second equation of (2.25) gives

$$\frac{d\bar{y}}{d\varepsilon} = 2(x + \varepsilon),$$

and integrating leads to

$$\bar{y} = 2x\varepsilon + \varepsilon^2 + c(x, y), \tag{2.27}$$

where again c is an arbitrary function of its arguments. Imposing the initial condition $\bar{y} = y$ when $\varepsilon = 0$ gives $c(x, y) = y$ gives

$$\bar{y} = y + 2x\varepsilon + \varepsilon^2.$$

■

EXAMPLE 2.7

Consider $X = x^2$ and $Y = y^2$. From (2.24) we are required to solve

$$\frac{d\bar{x}}{d\varepsilon} = \bar{x}^2, \quad \frac{d\bar{y}}{d\varepsilon} = \bar{y}^2, \tag{2.28}$$

subject to the initial conditions given in (2.24). The solution of the first is

$$-\frac{1}{\bar{x}} = \varepsilon + c(x, y),$$

and imposing the initial condition $\bar{x} = x$ when $\varepsilon = 0$ gives $c(x, y) = -1/x$. This leads to

$$-\frac{1}{\bar{x}} = \varepsilon - \frac{1}{x},$$

or

$$\bar{x} = \frac{x}{1 - \varepsilon x}.$$

As the second differential equation in (2.28) is the same form as the first, its solution is easily obtained, giving

$$\bar{y} = \frac{y}{1 - \varepsilon y}. \tag{2.29}$$

At this point we discover, if we had the infinitesimals X and Y, we could construct a transformation that would lead to a separable equation involving r and s. Consider

$$r = r(x, y), \quad s = s(x, y) \tag{2.30}$$

and require that (2.30) be invariant, that is,

$$\bar{r} = r(\bar{x}, \bar{y}), \quad \bar{s} = s(\bar{x}, \bar{y}). \tag{2.31}$$

Differentiating (2.31) with respect to ε gives

$$\frac{\partial \bar{r}}{\partial \varepsilon} = \frac{\partial r}{\partial \bar{x}}\frac{\partial \bar{x}}{\partial \varepsilon} + \frac{\partial r}{\partial \bar{y}}\frac{\partial \bar{y}}{\partial \varepsilon}, \quad \frac{\partial \bar{s}}{\partial \varepsilon} = \frac{\partial s}{\partial \bar{x}}\frac{\partial \bar{x}}{\partial \varepsilon} + \frac{\partial s}{\partial \bar{y}}\frac{\partial \bar{y}}{\partial \varepsilon}. \tag{2.32}$$

We also note from (2.5) that

$$\frac{\partial \bar{r}}{\partial \varepsilon} = 0, \quad \frac{\partial \bar{s}}{\partial \varepsilon} = 1. \tag{2.33}$$

Setting $\varepsilon = 0$ in (2.32) and using (2.33) and (2.21) gives

$$X(x, y)\frac{\partial r}{\partial x} + Y(x, y)\frac{\partial r}{\partial y} = 0, \quad X(x, y)\frac{\partial s}{\partial x} + Y(x, y)\frac{\partial s}{\partial y} = 1. \tag{2.34}$$

Thus, if we have the infinitesimals X and Y, solving (2.34) would give rise to the transformation that will separate the given ODE. ■

EXAMPLE 2.8

Recall that (2.2) left (2.1) invariant and we obtained the infinitesimals X and Y in (2.23). Then we need to solve

$$xr_x - yr_y = 0, \quad xs_x - ys_y = 1, \tag{2.35}$$

from (2.34). By the method of characteristics, we obtain the solutions of (2.35) as

$$r = R(xy), \quad s = \ln x + S(xy), \tag{2.36}$$

where R and S are arbitrary functions of their arguments. Choosing R and S in (2.36) such that $r = xy$ and $s = \ln x$ gives $x = e^s$ and $y = re^{-s}$, recovering the change of variables (2.3). ■

EXAMPLE 2.9

Recall that (2.7) left (2.6) invariant. The infinitesimals associated with (2.7) are $X = 1$ and $Y = y/x$. From (2.34), we need to solve

$$r_x + \frac{y}{x} r_y = 0, \quad s_x + \frac{y}{x} s_y = 1. \tag{2.37}$$

By the method of characteristics, we obtain the solution of (2.37) as

$$r = R\left(\frac{y}{x}\right), \quad s = x + S\left(\frac{y}{x}\right), \tag{2.38}$$

where R and S are arbitrary functions of their arguments. Choosing R and S in (2.38) such that $r = \frac{y}{x}$ and $s = x$ gives rise to the transformation (2.8). At this point, we are ready to answer the question: How do we find the infinitesimals X and Y? ∎

2.2 LIE'S INVARIANCE CONDITION

We have seen that given the infinitesimals X and Y, we can construct a change of variables leading to a separable equation; we now show how these infinitesimals are obtained. We wish to seek invariance of

$$\frac{\mathrm{d}y}{\mathrm{d}x} = F(x, y) \tag{2.39}$$

under the infinitesimal transformation

$$\bar{x} = x + X(x, y)\varepsilon + O(\varepsilon^2), \quad \bar{y} = y + Y(x, y)\varepsilon + O(\varepsilon^2). \tag{2.40}$$

In doing so, it is necessary to know how derivatives transform under the infinitesimal transformations (2.40). Under (2.40), we obtain

$$\frac{\mathrm{d}\bar{y}}{\mathrm{d}\bar{x}} = \frac{\frac{\mathrm{d}}{\mathrm{d}x}\left(y + Y(x, y)\varepsilon + O(\varepsilon^2)\right)}{\frac{\mathrm{d}}{\mathrm{d}x}\left(x + X(x, y)\varepsilon + O(\varepsilon^2)\right)}$$
$$= \frac{\frac{\mathrm{d}y}{\mathrm{d}x} + [Y_x + Y_y y']\varepsilon + O(\varepsilon^2)}{1 + [X_x + X_y y']\varepsilon + O(\varepsilon^2)}$$

$$= \left(\frac{\mathrm{d}y}{\mathrm{d}x} + [Y_x + Y_y y']\varepsilon + O(\varepsilon^2)\right)\left(1 - [X_x + X_y y']\varepsilon + O(\varepsilon^2)\right)$$

$$= \frac{\mathrm{d}y}{\mathrm{d}x} + \left(Y_x + [Y_y - X_x]y' - X_y y'^2\right)\varepsilon + O(\varepsilon^2). \tag{2.41}$$

We now consider the following ODE

$$\frac{\mathrm{d}\bar{y}}{\mathrm{d}\bar{x}} = F(\bar{x}, \bar{y}). \tag{2.42}$$

Substituting the infinitesimal transformations (2.40) and first-order derivative transformation (2.41) into (2.42) yields

$$\frac{\mathrm{d}y}{\mathrm{d}x} + \left(Y_x + [Y_y - X_x]y' - X_y y'^2\right)\varepsilon + O\left(\varepsilon^2\right)$$
$$= F\left(x + X(x,y)\varepsilon + O\left(\varepsilon^2\right), y + Y(x,y)\varepsilon + O\left(\varepsilon^2\right)\right).$$

Expanding to order $O(\varepsilon^2)$ gives

$$\frac{\mathrm{d}y}{\mathrm{d}x} + \left(Y_x + [Y_y - X_x]y' - X_y y'^2\right)\varepsilon + O\left(\varepsilon^2\right)$$
$$= F(x,y) + (XF_x + YF_y)\varepsilon + O\left(\varepsilon^2\right), \tag{2.43}$$

and imposing

$$\frac{\mathrm{d}y}{\mathrm{d}x} = F(x,y)$$

shows that equation (2.43) is satisfied to $O(\varepsilon^2)$ if

$$Y_x + \left(Y_y - X_x\right)F - X_y F^2 = XF_x + YF_y. \tag{2.44}$$

This is known as *Lie's Invariance Condition*. For a given $F(x,y)$, any functions $X(x,y)$ and $Y(x,y)$ that solve equation (2.44) are the infinitesimals that we seek.

Let's reexamine the preceding examples.

EXAMPLE 2.10

Consider

$$\frac{\mathrm{d}y}{\mathrm{d}x} = y^2 - \frac{y}{x} - \frac{1}{x^2},$$

which was found in Example 2.1 to be invariant under

$$\bar{x} = e^{\varepsilon} x, \quad \bar{y} = e^{-\varepsilon} y.$$

We found in Example 2.4 the corresponding infinitesimals to be

$$X = x, \quad Y = -y. \tag{2.45}$$

Lie's invariance condition (2.44) becomes

$$\begin{aligned} Y_x + (Y_y - X_x)\left(y^2 - \frac{y}{x} - \frac{1}{x^2}\right) - X_y\left(y^2 - \frac{y}{x} - \frac{1}{x^2}\right)^2 \\ = X\left(\frac{y}{x^2} + \frac{2}{x^3}\right) + Y\left(2y - \frac{1}{x}\right). \end{aligned} \tag{2.46}$$

Substituting the infinitesimals (2.45) into (2.46) gives

$$(-2)\left(y^2 - \frac{y}{x} - \frac{1}{x^2}\right) = x\left(\frac{y}{x^2} + \frac{2}{x^3}\right) - y\left(2y - \frac{1}{x}\right),$$

which, upon simplification shows, it's identically satisfied. Thus, $X(x, y) = x$ and $Y(x, y) = -y$ are solutions of Lie's invariance condition (2.46). ■

EXAMPLE 2.11

Consider

$$\frac{dy}{dx} = \frac{y}{x} + \frac{x^2}{x + y}.$$

We saw at the beginning of this chapter that this ODE is invariant under

$$\bar{x} = x + \varepsilon, \quad \bar{y} = \frac{(x + \varepsilon) y}{x},$$

with the associated infinitesimals

$$X = 1, \quad Y = \frac{y}{x}. \tag{2.47}$$

Lie's invariance condition becomes

$$\begin{aligned} Y_x + (Y_y - X_x)\left(\frac{y}{x} + \frac{x^2}{x + y}\right) - X_y\left(\frac{y}{x} + \frac{x^2}{x + y}\right)^2 \\ = X\left(-\frac{y}{x^2} + \frac{x^2 + 2xy}{(x + y)^2}\right) + Y\left(\frac{1}{x} - \frac{x^2}{(x + y)^2}\right). \end{aligned} \tag{2.48}$$

The reader can verify that X and Y given in (2.47) satisfies (2.48) but instead we will try and deduce these particular X and Y. As (2.48) is difficult to solve for $X(x, y)$ and $Y(x, y)$ in general, we will seek special solutions of the form

$$X = A(x), \quad Y = B(x)y. \tag{2.49}$$

Substituting (2.49) into equation (2.48), simplifying and collecting coefficients of the numerator with respect to y gives

$$\begin{aligned} &\left(A - A'x + B'x^2\right) y^3 + 2\left(A - A'x + B'x^2\right) xy^2 \\ &+ \left(A - (2A + A')x + (2B - A' + B')x^2\right) x^2 y - (A - Bx + A'x)x^4 = 0. \end{aligned} \tag{2.50}$$

As this must be satisfied for all y, then the coefficients must be identically zero, giving rise to the following equations for A and B:

$$A - A'x + B'x^2 = 0, \tag{2.51a}$$

$$A - (2A + A')x + (2B - A' + B')x^2 = 0, \tag{2.51b}$$

$$-A + Bx - A'x = 0. \tag{2.51c}$$

Solving (2.51c) for B gives

$$B = A' + \frac{A}{x} \tag{2.52}$$

and substituting into the remaining equations of (2.51) gives

$$A'' = 0, \quad A'' + A' = 0,$$

from which it is easy to deduce that $A = c_1$. From (2.52), $B = \dfrac{c}{x}$, where c is an arbitrary constant. Thus, from (2.49), we find the infinitesimals X and Y to be

$$X = c, \quad Y = \frac{cx}{y},$$

and setting $c = 1$ gives exactly those infinitesimals given in (2.47).

In finding the infinitesimals X and Y in (2.48), it was necessary to guess a simplified form, namely (2.49). This is often the case. We usually try the following forms (see the exercises):

$$\begin{aligned} &\text{(i)} \quad X = A(x), \quad Y = B(x)y + C(x), \\ &\text{(ii)} \quad X = A(y), \quad Y = B(y)x + C(y), \\ &\text{(iii)} \quad X = A(x), \quad Y = B(y), \\ &\text{(iv)} \quad X = A(y), \quad Y = B(x). \end{aligned}$$

■

EXERCISES

1. For the given Lie group, find the corresponding infinitesimals X and Y.

$$\text{(i)}\quad \bar{x} = \cosh\varepsilon x + \sinh\varepsilon y, \quad \bar{y} = \cosh\varepsilon y + \sinh\varepsilon x$$

$$\text{(ii)}\quad \bar{x} = \sqrt{x^2 - 2y + 2ye^{\varepsilon}}, \quad \bar{y} = ye^{\varepsilon},$$

$$\text{(iii)}\quad \bar{x} = \frac{x}{\sqrt{1+2\varepsilon y^2}}, \qquad \bar{y} = \frac{y}{\sqrt{1+2\varepsilon y^2}}.$$

2. For the given infinitesimals X and Y, find the corresponding Lie group.

$$\text{(i)}\quad X = xy, \quad Y = 1$$

$$\text{(ii)}\quad X = x + y, \quad Y = y,$$

$$\text{(iii)}\quad X = x^2 - x, \quad Y = y^2 + y.$$

3. Find infinitesimal transformations leaving the following ODEs invariant. Use these to find a change of variables and reduce the original ODE to one that is separable and solve the equation.

$$(i)\quad \frac{dy}{dx} = xy + \frac{y}{x} + \frac{e^{x^2}}{xy}$$

$$(ii)\quad \frac{dy}{dx} = \frac{3y}{x} + \frac{x^5}{2y + x^3}$$

$$(iii)\quad \frac{dy}{dx} = \frac{y + y^3}{x + (x+1)y^2}$$

Hints: Try:
1. $X = A(x),\ Y = B(x)y,$
2. $X = A(x),\ Y = B(y),$
3. $X = A(y),\ Y = B(x).$

4*. If we introduce the change of variables given in (2.34), namely

$$X\frac{\partial r}{\partial x} + Y\frac{\partial r}{\partial y} = 0, \quad \frac{\partial s}{\partial x} + Y\frac{\partial s}{\partial y} = 1,$$

such that the differential equation

$$\frac{dy}{dx} = F(x, y)$$

is transformed to one that is independent of s, prove that X, Y, and F satisfy Lie' invariance condition (2.44).

2.3 STANDARD INTEGRATION TECHNIQUES

When one first learns the standard techniques for solving first-order ordinary differential equations, one creates a cookbook on how to solve a wide variety of equations: *linear*, *Bernoulli*, *homogeneous*, *exact*, *Riccati*, *etc.* A natural question would be: Do these equations have anything in common? The answer is yes—they are all invariant under some infinitesimal transformation.

2.3.1 Linear Equations

The general form of a linear ODE is

$$\frac{\mathrm{d}y}{\mathrm{d}x} + P(x)y = Q(x). \tag{2.53}$$

This is invariant under the Lie group

$$\bar{x} = x, \quad \bar{y} = y + \varepsilon \mathrm{e}^{-\int P(x)\mathrm{d}x} \tag{2.54}$$

as

$$\frac{\mathrm{d}\bar{y}}{\mathrm{d}\bar{x}} = \frac{\mathrm{d}y}{\mathrm{d}x} + \varepsilon \left(-P(x)\mathrm{e}^{-\int P(x)\mathrm{d}x}\right)$$

and

$$\frac{\mathrm{d}\bar{y}}{\mathrm{d}\bar{x}} + P(\bar{x})\bar{y} = Q(\bar{x})$$

gives

$$\frac{\mathrm{d}y}{\mathrm{d}x} + \varepsilon \left(-P(x)\mathrm{e}^{-\int P(x)\mathrm{d}x}\right) + P(x)\left(y + \varepsilon \mathrm{e}^{-\int P(x)\mathrm{d}x}\right) = Q(x). \tag{2.55}$$

Expanding (2.55) gives

$$\frac{\mathrm{d}y}{\mathrm{d}x} + P(x)y = Q(x).$$

From the Lie group (2.54), we obtain the infinitesimals $X = 0$ and $Y = \mathrm{e}^{-\int P(x)\mathrm{d}x}$. The change of variables are obtained by solving (2.34), so

$$\mathrm{e}^{-\int P(x)\mathrm{d}x} r_y = 0, \quad \mathrm{e}^{-\int P(x)\mathrm{d}x} s_y = 1. \tag{2.56}$$

Thus,

$$\begin{aligned} r_y = 0 \quad &\Rightarrow \quad r = R(x), \\ s_y = \mathrm{e}^{\int P(x)\mathrm{d}x} \quad &\Rightarrow \quad s = \mathrm{e}^{\int P(x)\mathrm{d}x} y + S(x), \end{aligned}$$

where $R(x)$ and $S(x)$ are arbitrary functions. Choosing $R = x$ and $S = 0$ gives

$$x = r, \quad y = \mathrm{e}^{-\int P(r)\mathrm{d}r} s.$$

Calculating $\dfrac{\mathrm{d}y}{\mathrm{d}x}$ gives

$$\frac{\mathrm{d}y}{\mathrm{d}x} = \mathrm{e}^{-\int P(r)\mathrm{d}r} \frac{\mathrm{d}s}{\mathrm{d}r} - P(r)\mathrm{e}^{-\int P(r)\mathrm{d}r} s$$

and substituting into (2.53) and simplifying gives

$$\frac{\mathrm{d}s}{\mathrm{d}r} = \mathrm{e}^{\int P(r)\mathrm{d}r} Q(r),$$

which is a separable equation.

EXAMPLE 2.12

Consider

$$y' + y = x. \tag{2.57}$$

Here $P(x) = 1$, so

$$Y = \mathrm{e}^{-\int P(x)\mathrm{d}x} = \mathrm{e}^{-x}.$$

Thus, from (2.56), we need to solve

$$r_y = 0, \quad \mathrm{e}^{-x} s_y = 1.$$

This gives

$$r = R(x), \quad s = \mathrm{e}^{x} y + S(x),$$

where R and S are arbitrary, and choosing $R(x) = x$ and $S(x) = 0$ gives

$$r = x, \quad s = e^{x}y.$$

Under this change of variables, the linear equation (2.57) becomes

$$\frac{ds}{dr} = re^{r},$$

which is separable. ■

2.3.2 Bernoulli Equation

Equations of the form

$$\frac{dy}{dx} + P(x)y = Q(x)y^{n}, \quad n \neq 0, 1$$

are called Bernoulli equations. Assuming $X = 0$, Lie's invariance condition (2.44) becomes

$$Y_x + Y_y\left(Q(x)y^n - P(x)y\right) = Y\left(nQ(x)y^{n-1} - P(x)\right). \tag{2.58}$$

The reader can verify that

$$Y = e^{(n-1)\int P(x)dx}y^{n}$$

satisfies (2.58). To obtain a change of variables, it is necessary to solve

$$e^{(n-1)\int P(x)dx}y^{n}r_y = 0, \quad e^{(n-1)\int P(x)dx}y^{n}s_y = 1. \tag{2.59}$$

We best show this via an example.

EXAMPLE 2.13

Consider

$$y' + \frac{y}{x} = xy^3. \tag{2.60}$$

Here $P(x) = \dfrac{1}{x}$ and $n = 3$, so

$$Y = e^{(n-1)\int P(x)dx}y^{n} = e^{2\ln x}y^{3} = x^{2}y^{3}.$$

We need to solve (2.59), that is,

$$r_y = 0, \quad x^2y^3s_y = 1.$$

This gives

$$r = R(x), \quad s = S(x) + \frac{-1}{2x^2y^2},$$

and choosing $R = x$ and $S = 0$ gives

$$r = x, \quad s = \frac{-1}{2x^2y^2}.$$

The Bernoulli equation (2.60) then becomes

$$\frac{\mathrm{d}s}{\mathrm{d}r} = -\frac{1}{4r},$$

which is separable. ∎

2.3.3 Homogeneous Equations

Equations that are homogeneous are of the form

$$\frac{\mathrm{d}y}{\mathrm{d}x} = F\left(\frac{y}{x}\right). \tag{2.61}$$

Equation (2.61) is invariant under the Lie group

$$\overline{x} = \mathrm{e}^{\varepsilon}x, \quad \overline{y} = \mathrm{e}^{\varepsilon}y, \tag{2.62}$$

and expanding (2.62) gives

$$\overline{x} = x + \varepsilon x + \mathrm{O}(\varepsilon^2),$$
$$\overline{y} = y + \varepsilon y + \mathrm{O}(\varepsilon^2),$$

giving the infinitesimals $X = x$ and $Y = y$. Therefore, the change of variables is obtained by solving

$$xr_x + yr_y = 0, \quad xs_x + ys_y = 1,$$

which has solutions

$$r = R\left(\frac{y}{x}\right), \quad s = \ln x + S\left(\frac{y}{x}\right), \tag{2.63}$$

with R and S arbitrary functions of their arguments. Choosing R and S such that

$$r = \frac{y}{x}, \quad s = \ln x$$

gives

$$x = \mathrm{e}^s \text{ and } y = \mathrm{e}^s r \tag{2.64}$$

and

$$\frac{\mathrm{d}y}{\mathrm{d}x} = \frac{\dfrac{\mathrm{d}}{\mathrm{d}r}(r\mathrm{e}^s)}{\dfrac{\mathrm{d}}{\mathrm{d}r}\mathrm{e}^s} = \frac{\mathrm{e}^s + r\mathrm{e}^s s'}{\mathrm{e}^s s'} = \frac{1 + rs'}{s'}. \tag{2.65}$$

By substituting (2.64) and (2.65) into (2.61), we obtain

$$\frac{1 + rs'}{s'} = F(r),$$

or

$$s' = \frac{1}{F(r) - r},$$

a separable ODE. We further note that in the case $F(r) = r$, the original equation is separable!

EXAMPLE 2.14

Consider

$$y' = \frac{y + \sqrt{x^2 + y^2}}{x}. \tag{2.66}$$

As this ODE is invariant under (2.62), we use the change of variables given in (2.64). In doing so, our homogeneous ODE (2.66) becomes

$$\frac{\mathrm{d}s}{\mathrm{d}r} = \frac{1}{\sqrt{r^2 + 1}},$$

which is separable. ■

2.3.4 Exact Equations

The equation

$$\frac{\mathrm{d}y}{\mathrm{d}x} = F(x, y) \tag{2.67}$$

can be rewritten as

$$F(x, y)\mathrm{d}x - \mathrm{d}y = 0$$

or

$$M(x, y)\mathrm{d}x + N(x, y)\mathrm{d}y = 0. \tag{2.68}$$

Equation (2.68) is said to be exact if

$$\frac{\partial M}{\partial y} = \frac{\partial N}{\partial x}. \tag{2.69}$$

If (2.69) holds, there exists a function ϕ such that

$$\frac{\partial \phi}{\partial x} = M, \quad \frac{\partial \phi}{\partial y} = N,$$

whose solution is

$$\phi(x, y) = C,$$

which is the solution of the ODE (2.68). This is rarely the case. Sometimes, we look for an integrating factor μ such that

$$\mu M\mathrm{d}x + \mu N\mathrm{d}y = 0$$

is exact. That is,

$$\frac{\partial}{\partial y}(\mu M) = \frac{\partial}{\partial x}(\mu N). \tag{2.70}$$

This is still a hard problem, as we now have a PDE for μ. Suppose that we knew that equation (2.68) was invariant under the infinitesimal transformation

$$\begin{aligned} \overline{x} &= x + \varepsilon X(x, y) + \mathrm{O}(\varepsilon^2) \\ \overline{y} &= y + \varepsilon Y(x, y) + \mathrm{O}(\varepsilon^2), \end{aligned}$$

then an integrating factor for (2.68) is given by

$$\mu = \frac{1}{MX + NY}. \tag{2.71}$$

This result is actually due to Lie himself [2].

From (2.70) with μ given in (2.71), we get

$$\frac{\partial}{\partial y}\left(\frac{M}{MX+NY}\right)=\frac{\partial}{\partial x}\left(\frac{N}{MX+NY}\right). \tag{2.72}$$

From the original ODE (2.67) we can identify that $M=-FN$ and substituting into (2.72) gives

$$\frac{\partial}{\partial y}\left(\frac{-F}{Y-FX}\right)=\frac{\partial}{\partial x}\left(\frac{1}{Y-FX}\right).$$

Expanding gives

$$\frac{-F_y(Y-FX)+F(Y_y-F_yX-FX_y)}{(Y-FX)^2}=-\frac{Y_x-F_xX-FX_x}{(Y-FX)^2}$$

and simplifying gives

$$Y_x+(Y_y-X_x)F-X_yF^2=XF_x+YF_y$$

which is exactly Lie's invariance condition (2.44).

EXAMPLE 2.15

Consider

$$\left(2x^4y+y^4\right)\mathrm{d}x+\left(x^5-2xy^3\right)\mathrm{d}y=0. \tag{2.73}$$

Here M and N are given by

$$M=2x^4+y^4,\quad N=x^5-2xy^3,$$

so

$$\frac{\partial M}{\partial y}=2x^4+4y^3,\quad \frac{\partial N}{\partial x}=5x^4-2y^3.$$

These are clearly not equal so the ODE in not exact. Equation (2.73) is invariant under

$$\overline{x}=\mathrm{e}^{3\varepsilon}x,\quad \overline{y}=\mathrm{e}^{4\varepsilon}y,$$

so we obtain the infinitesimals X and Y as

$$X=3x,\quad Y=4y.$$

From (2.71), we obtain μ as

$$\mu = \frac{1}{3x(2x^4y + y^4) + 4y(x^5 - 2xy^3)} = \frac{1}{5xy(2x^4 - y^3)}.$$

Therefore,

$$\frac{\partial \phi}{\partial x} = \mu M = \frac{y(2x^4 + y^3)}{5xy(2x^4 - y^3)}, \quad \frac{\partial \phi}{\partial y} = \mu N = \frac{x(x^4 - 2y^3)}{5xy(2x^4 - y^3)}. \tag{2.74}$$

Note that the constant 5 will be omitted as this will not change the final result. Integrating the first equation of (2.74) gives

$$\phi = -\ln|x| + \frac{1}{2}\ln|2x^4 - y^3| + F(y),$$

where F is an arbitrary function of integration. Substitution into the second equation of (2.74) and expanding gives

$$F'(y) = \frac{1}{2y} \quad \Rightarrow \quad F(y) = \frac{1}{2}\ln|y| + c.$$

Therefore,

$$\phi = -\ln|x| + \frac{1}{2}\ln|2x^4 - y^3| + \frac{1}{2}\ln|y| + c.$$

The solution of the ODE (2.73) is therefore given by $\phi = \phi_0$ (constant) and after simplification gives

$$\frac{(2x^4 - y^3)y}{x^2} = k,$$

where k is an arbitrary constant. ■

2.3.5 Riccati Equations

The general form of a Riccati ODE is

$$\frac{\mathrm{d}y}{\mathrm{d}x} = P(x)y^2 + Q(x)y + R(x). \tag{2.75}$$

Our goal is to find X and Y that satisfies Lie's invariance condition (2.44). As with both linear and Bernoulli equations, we will assume $X = 0$, giving Lie's invariance condition as

$$Y_x + \left(P(x)y^2 + Q(x)y + R(x)\right) Y_y = (2P(x)y + Q(x))\, Y. \tag{2.76}$$

One solution of (2.76) is

$$Y = (y - y_1)^2 F(x),$$

where y_1 is one solution to (2.75) and F satisfies

$$F' + (2Py_1 + Q) F = 0. \tag{2.77}$$

In order to solve for the canonical variables r and s, it is necessary to solve

$$(y - y_1)^2 F(x) r_y = 0, \quad (y - y_1)^2 F(x) s_y = 1,$$

from which we obtain

$$r = R(x), \quad s = S(x) - \frac{1}{(y - y_1)F},$$

where R and S are arbitrary functions. Setting $R(x) = x$ and $S(x) = 0$ yields

$$x = r, \quad y = y_1 - \frac{1}{sF(r)}, \tag{2.78}$$

thereby transforming the original Riccati equation (2.75) to

$$\frac{ds}{dr} = \frac{a(r)}{F(r)}.$$

It is interesting that the usual linearizing transformation is recovered using Lie's method.

EXAMPLE 2.16

Consider

$$y' = \frac{y^2}{e^x} + 2y - 2e^x.$$

One solution of this Riccati equation is $y_1 = e^x$. From (2.77), we see that F satisfies

$$F' + 4F = 0,$$

from which we obtain the solution $F = e^{-4x}$, noting that the constant of integration can be set to one without the loss of generality. Thus, under the change of variables given in (2.78), namely

$$x = r, \quad y = e^r - \frac{e^{4r}}{s}, \tag{2.79}$$

the original ODE becomes

$$\frac{ds}{dr} = e^{3r}.$$

■

EXERCISES

1. Find infinitesimal transformations leaving the following ODEs invariant and use these to separate the following

$$\text{(i)}\ \frac{dy}{dx} = \frac{e^x}{x^3} - \frac{3y}{x}, \qquad \text{(ii)}\ \frac{dy}{dx} = 2y + x$$

$$\text{(iii)}\ \frac{dy}{dx} = 2y + xy^2, \qquad \text{(iv)}\ \frac{dy}{dx} = \frac{x^3}{y} - xy,$$

$$\text{(v)}\ \frac{dy}{dx} = \frac{x^3 + y^3}{x^2 y}, \qquad \text{(vi)}\ \frac{dy}{dx} = \ln y - \ln x,$$

$$\text{(vii)}\ \frac{dy}{dx} = \frac{y - 2xy^2}{x + 2x^2 y}, \qquad \text{(viii)}\ \frac{dy}{dx} = -\frac{3xy + 2y^3}{x^2 + 3xy^2},$$

$$\text{(ix)}\ \frac{dy}{dx} = \frac{y^2}{x^2} - 2\frac{y}{x} + 2, \qquad \text{(x)}\ \frac{dy}{dx} = e^x y^2 + y - 3e^{-x},$$

$$y_1 = x \qquad\qquad y_1 = e^{-x}.$$

2. Consider

$$\left(2xy^2 + 3x^2\right) dx + \left(2x^2 y - 4\right) dy = 0,$$

which is already in exact form as $M_y = N_x$. Show from (2.71) that X and Y satisfy

$$\left(2xy^2 + 3x^2\right) X + \left(2x^2 y - 4\right) Y = 1.$$

Further show that this satisfies Lies' invariance condition (2.44). This illustrates that first-order ODEs have an infinite number of symmetries.

3. Consider the ODE

$$2\,y\,\mathrm{d}x + \left(x + 5y^2\right)\mathrm{d}y = 0. \tag{2.80}$$

Show Lie's invariance condition (2.44) for this ODE admits the solutions

$$X = 2x, \quad Y = y, \qquad \text{and} \qquad X = \frac{x}{y^2\sqrt{y}}, \quad Y = -\frac{2}{y\sqrt{y}}.$$

Further show from (2.71) that the integrating factors are (to within a multiplicative constant)

$$\mu = \frac{1}{xy + y^3}, \quad \text{and} \quad \mu = \frac{1}{\sqrt{y}},$$

thus illustrating that integrating factors are not unique.

2.4 INFINITESIMAL OPERATOR AND HIGHER ORDER EQUATIONS

So far, we have considered infinitesimal transformations of the form

$$\overline{x} = x + \varepsilon X(x, y) + \mathrm{O}(\varepsilon^2) \tag{2.81a}$$

$$\overline{y} = y + \varepsilon Y(x, y) + \mathrm{O}(\varepsilon^2) \tag{2.81b}$$

with

$$\frac{\mathrm{d}\overline{y}}{\mathrm{d}\overline{x}} = \frac{\mathrm{d}y}{\mathrm{d}x} + \varepsilon\left[Y_x + \left(Y_y - X_x\right)y' - X_y y'^2\right] + \mathrm{O}(\varepsilon^2). \tag{2.82}$$

In order to extend the method to higher order equations, it is useful to introduce a more compact notation.

2.4.1 The Infinitesimal Operator

We define the infinitesimal operator as

$$\Gamma = X\frac{\partial}{\partial x} + Y\frac{\partial}{\partial y}. \tag{2.83}$$

As an example of the usefulness of this notation, consider $F(\overline{x}, \overline{y})$. We can write

$$\begin{aligned} F(\overline{x}, \overline{y}) &= F\left(x + \varepsilon X + O(\varepsilon^2), y + \varepsilon Y + O(\varepsilon^2)\right), \\ &= F(x, y) + \varepsilon\left(XF_x + YF_y\right) + O(\varepsilon^2), \\ &= F(x, y) + \varepsilon \Gamma F + O(\varepsilon^2), \end{aligned}$$

where

$$\Gamma F = \left(X\frac{\partial}{\partial x} + Y\frac{\partial}{\partial y}\right)F.$$

2.4.2 The Extended Operator

The invariance of $\dfrac{dy}{dx} = F(x, y)$ leads to

$$Y_x + \left(Y_y - X_x\right)y' - X_y y'^2 = XF_x + YF_y. \tag{2.84}$$

As we have introduced an infinitesimal operator in (2.83), we now introduce the extended operator

$$\Gamma^{(1)} = X\frac{\partial}{\partial x} + Y\frac{\partial}{\partial y} + Y_{[x]}\frac{\partial}{\partial y'}. \tag{2.85}$$

If we define Δ such that

$$\Delta = \frac{dy}{dx} - F(x, y) = 0,$$

then

$$\Gamma^{(1)}\Delta = Y_{[x]} - XF_x - YF_y = 0. \tag{2.86}$$

Comparing (2.84) and (2.86) shows that we can define $Y_{[x]}$ as

$$Y_{[x]} = Y_x + \left(Y_y - X_x\right)y' - X_y y'^2.$$

When we substitute $y' = F$ into

$$\Gamma^{(1)}\Delta = 0,$$

we get Lie's invariance condition. Thus,

$$\Gamma^{(1)}\Delta\Big|_{\Delta=0} = 0$$

is a convenient way to write the invariance condition.

2.4.3 Extension to Higher Orders

It is now time to reexamine the infinitesimal $Y_{[x]}$. From (2.82) and 2.87, we see that we can write

$$\frac{\mathrm{d}\bar{y}}{\mathrm{d}\bar{x}} = \frac{\mathrm{d}y}{\mathrm{d}x} + Y_{[x]}\varepsilon + \mathrm{O}(\varepsilon^2),$$

where

$$Y_{[x]} = Y_x + \left(Y_y - X_x\right)y' - X_y y'^2.$$

This suggests that we could write higher derivatives as

$$\frac{\mathrm{d}^2\bar{y}}{\mathrm{d}\bar{x}^2} = \frac{\mathrm{d}^2 y}{\mathrm{d}x^2} + Y_{[xx]}\varepsilon + \mathrm{O}(\varepsilon^2),$$

and

$$\frac{\mathrm{d}^3\bar{y}}{\mathrm{d}\bar{x}^3} = \frac{\mathrm{d}^3 y}{\mathrm{d}x^3} + Y_{[xxx]}\varepsilon + \mathrm{O}(\varepsilon^2),$$

and so on.

2.4.4 First-Order Infinitesimals (revisited)

Before we look at the extension to higher orders, consider

$$\begin{aligned} Y_{[x]} &= Y_x + Y_y y' - X_x y' - X_y y'^2 \\ &= Y_x + Y_y y' - (X_x + X_y y')y'. \end{aligned} \tag{2.87}$$

If we define the total differential operator as

$$D_x = \frac{\partial}{\partial x} + y'\frac{\partial}{\partial y},$$

then

$$Y_{[x]} = D_x(Y) - D_x(X)y'.$$

We call $Y_{[x]}$ an extended infinitesimal .

Consider $\frac{d\overline{y}}{d\overline{x}}$ using the total differential operator D. Therefore,

$$
\begin{aligned}
\frac{d\overline{y}}{d\overline{x}} &= \frac{\frac{d}{dx}\left(y + \varepsilon Y + O(\varepsilon^2)\right)}{\frac{d}{dx}\left(x + \varepsilon X + O(\varepsilon^2)\right)} \\
&= \frac{\frac{dy}{dx} + \varepsilon D_x(Y) + O(\varepsilon^2)}{1 + \varepsilon D_x(X) + O(\varepsilon^2)} \\
&= \left(\frac{dy}{dx} + \varepsilon D_x(Y) + O(\varepsilon^2)\right)\left(1 - \varepsilon D_x(X) + O(\varepsilon^2)\right) \\
&= \frac{dy}{dx} + \left(D_x(Y) - D_x(X)y'\right) + O(\varepsilon^2) \\
&= \frac{dy}{dx} + Y_{[x]}\varepsilon + O(\varepsilon^2).
\end{aligned}
$$

2.4.5 Second-Order Infinitesimals

We now consider second-order extended infinitesimals:

$$
\begin{aligned}
\frac{d^2\overline{y}}{d\overline{x}^2} &= \frac{d}{d\overline{x}}\left(\frac{d\overline{y}}{d\overline{x}}\right) \\
&= \frac{d}{dx}\left(\frac{d\overline{y}}{d\overline{x}}\right)\Big/\frac{d\overline{x}}{dx} \\
&= \frac{\frac{d}{dx}\left[\frac{dy}{dx} + \left[D_x(Y) - D_x(X)y'\right]\varepsilon + O(\varepsilon^2)\right]}{1 + \varepsilon D_x(X) + O(\varepsilon^2)} \\
&= \left(\frac{d^2y}{dx^2} + D_x(Y_{[x]})\varepsilon + O(\varepsilon^2)\right)\left(1 - \varepsilon D_x(X) + O(\varepsilon^2)\right) \\
&= \frac{d^2y}{dx^2} + \left(D_x(Y_{[x]}) - D_x(X)y''\right)\varepsilon + O(\varepsilon^2),
\end{aligned}
$$

so

$$
Y_{[xx]} = D_x(Y_{[x]}) - y'' D_x(X).
$$

As $Y_{[x]}$ contains x, y, and y', we need to extend the definition of D_x, so

$$D_x = \frac{\partial}{\partial x} + y'\frac{\partial}{\partial y} + y''\frac{\partial}{\partial y'} + y'''\frac{\partial}{\partial y''} + \cdots$$

and expanding $Y_{[xx]}$ gives

$$\begin{aligned} Y_{[xx]} &= D_x(Y_x + (Y_y - X_x)y' - X_y y'^2) - y'' D_x(X) \\ &= Y_{xx} + (2Y_{xy} - X_{xx})y' + (Y_{yy} - 2X_{xy})y' - X_{yy}y'^3, \\ &-(Y_y - 2X_x)y'' - 3X_y y' y''. \end{aligned}$$

2.4.6 The Invariance of Second-Order Equations

We consider

$$\frac{d^2\overline{y}}{d\overline{x}^2} = F\left(\overline{x}, \overline{y}, \frac{d\overline{y}}{d\overline{x}}\right),$$

so

$$\frac{d^2y}{dx^2} + \varepsilon Y_{[xx]} + O(\varepsilon^2) = F\big(x + \varepsilon X + O(\varepsilon^2), y + \varepsilon Y + O(\varepsilon^2), \frac{dy}{dx} + \varepsilon Y_{[x]} + O(\varepsilon^2)\big),$$

that, upon expansion, gives

$$\frac{d^2y}{dx^2} + \varepsilon Y_{[xx]} + O(\varepsilon^2) = F\big(x, y, \frac{dy}{dx}\big) + \varepsilon(XF_x + YF_y + Y_{[x]}F_{y'}) + O(\varepsilon^2). \quad (2.88)$$

If

$$\frac{d^2y}{dx^2} = F\left(x, y, \frac{dy}{dx}\right)$$

then (2.88) becomes

$$Y_{[xx]} = XF_x + YF_y + Y_{[x]}F_{y'}. \quad (2.89)$$

In terms of the extended operator $\Gamma^{(1)}$, (2.89) becomes

$$Y_{[xx]} = \Gamma^{(1)}F.$$

It is therefore natural to define a second extension to Γ:

$$\Gamma^{(2)} = X\frac{\partial}{\partial x} + Y\frac{\partial}{\partial y} + Y_{[x]}\frac{\partial}{\partial y'} + Y_{[xx]}\frac{\partial}{\partial y''}.$$

If we denote our ODE as Δ,

$$\Delta = y'' - f(x, y, y'),$$

then the invariance condition is

$$\Gamma^{(2)}\Delta\Big|_{\Delta=0} = 0.$$

2.4.7 Equations of arbitrary order

In general, if we define the nth order extension to Γ as

$$\Gamma^{(n)} = X\frac{\partial}{\partial x} + Y\frac{\partial}{\partial y} + Y_{[x]}\frac{\partial}{\partial y'} + Y_{[xx]}\frac{\partial}{\partial y''} + \cdots + Y_{[nx]}\frac{\partial}{\partial y^{(n)}} + \cdots,$$

where the infinitesimals are given by

$$Y_{[(n)x]} = D_x\left(Y_{[(n-1)x]}\right) - y^{(n)}D_x(X).$$

Then, invariance of the differential equation

$$\Delta\left(x, y, y', \ldots, y^{(n)}\right) = 0$$

is given by

$$\Gamma^{(n)}\Delta\Big|_{\Delta=0} = 0.$$

2.5 SECOND-ORDER EQUATIONS

In this section, we consider three examples to demonstrate the procedure of Lie's invariance condition for second-order ODEs. The first example is $y'' = 0$. As this ODE is trivial to integrate, we view this example solely to understand the algorithmic nature of Lie's method.

EXAMPLE 2.17

Consider the following ODE

$$y'' = 0.$$

If we denote this equation by

$$\Delta = y'' = 0,$$

then Lie's invariance condition is

$$\begin{aligned} \Gamma^{(2)}\Delta\Big|_{\Delta=0} &= 0, \\ \Gamma^{(2)}(y'')\Big|_{y''=0} &= 0, \\ Y_{[xx]}\Big|_{y''=0} &= 0. \end{aligned} \tag{2.90}$$

Substituting the extended infinitesimals transformations

$$\begin{aligned} Y_{[x]} &= D_x(Y) - y'D_x(X) \\ Y_{[xx]} &= D_x(Y_{[x]}) - y''D_x(X) \end{aligned}$$

into equation (2.90) and expanding gives

$$\begin{aligned} Y_{xx} + (2Y_{xy} - X_{xx})y' + (Y_{yy} - 2X_{xy})y'^2 - X_{yy}y'^3 \\ + (Y_y - 2X_x)y'' - 3X_y y' y'' = 0. \end{aligned} \tag{2.91}$$

Substituting $y'' = 0$ into equation (2.91) gives

$$Y_{xx} + (2Y_{xy} - X_{xx})y' + (Y_{yy} - 2X_{xy})y'^2 - X_{yy}y'^3 = 0.$$

When we set the coefficients of y', y'^2, and y'^3 to zero in (2.17), we get

$$Y_{xx} = 0, \tag{2.92a}$$

$$2Y_{xy} - X_{xx} = 0, \tag{2.92b}$$

$$Y_{yy} - 2X_{xy} = 0, \tag{2.92c}$$

$$X_{yy} = 0, \tag{2.92d}$$

an over-determined system of equations for the unknowns X and Y. Integrating equation (2.92d) gives

$$X(x, y) = a(x)y + b(x), \tag{2.93}$$

where a and b are arbitrary functions. Substituting into (2.92c) and integrating gives

$$\begin{aligned} Y_{yy} &= 2X_{xy} = 2a'(x) \\ Y_y &= 2a'(x)y + p(x) \\ Y &= a'(x)y^2 + p(x)y + q(x), \end{aligned} \tag{2.94}$$

where p and q are further arbitrary functions. Substituting (2.93) and (2.94) into equation (2.92b) gives

$$4a''(x)y + 2p'(x) - a''(x)y - b''(x) = 0.$$

When we set the coefficients of y to zero, we obtain

$$a''(x) = 0, \tag{2.95a}$$

$$2p'(x) - b''(x) = 0. \tag{2.95b}$$

Substituting (2.93) and (2.94) into equation (2.92a) gives

$$a'''(x)y^2 + p''(x)y + q''(x) = 0,$$

from which we obtain

$$p''(x) = 0, \tag{2.96a}$$

$$q''(x) = 0. \tag{2.96b}$$

The solutions of (2.95) and (2.96) are given simply as

$$\begin{aligned} a(x) &= a_1 x + a_2, \\ p(x) &= p_1 x + p_2, \\ q(x) &= q_1 x + q_2, \end{aligned}$$

where a_1, a_2, p_1, p_2, q_1, and q_2 are constant. The remaining function $b(x)$ is obtained from equation (2.95b), that is,

$$b''(x) = 2p_1,$$
$$b'(x) = 2p_1 x + b_1,$$
$$b(x) = p_1 x^2 + b_1 x + b_2,$$

where b_1 and b_2 are further constants. This leads to the infinitesimals X and Y as

$$X = (a_1 x + a_2)y + p_1 x^2 + b_1 x + b_2,$$
$$Y = a_1 y^2 + (p_1 x + p_2)y + q_1 x + q_2.$$

■

EXAMPLE 2.18

Consider the following ODE

$$y'' + yy' + xy^4 = 0. \tag{2.97}$$

If we denote (2.97) by

$$\Delta = y'' + yy' + xy^4 = 0, \tag{2.98}$$

then Lie's invariance condition is

$$\Gamma^{(2)}\Delta\Big|_{\Delta=0} = 0,$$

or in terms of the extended infinitesimals

$$Y_{[xx]} + Yy' + yY_{[x]} + Xy^4 + 4xy^3 Y = 0. \tag{2.99}$$

Substituting the extended infinitesimal transformations

$$Y_{[x]} = D_x(Y) - y' D_x(X)$$
$$Y_{[xx]} = D_x(Y_{[x]}) - y'' D_x(X)$$

into equation (2.99) and expanding gives

$$\begin{aligned} &Y_{xx} + (2Y_{xy} - X_{xx})y' + (Y_{yy} - 2X_{xy})y'^2 - X_{yy}y'^3 \\ &+ (Y_y - 2X_x)y'' - 3X_y y' y'' + Yy' \\ &+ y[Y_x + (Y_y - X_x)y' - X_y y'^2] + Xy^4 + 4xy^3 Y = 0. \end{aligned} \tag{2.100}$$

Substituting $y'' = -yy' - xy^4$ into equation (2.100) gives

$$\begin{aligned} &Y_{xx} + (2Y_{xy} - X_{xx})y' + (Y_{yy} - 2X_{xy})y'^2 - X_{yy}y'^3 \\ &\quad + (Y_y - 2X_x)(-yy' - xy^4) - 3X_y y'(-yy' - xy^4) + Yy' \\ &\quad + y[Y_x + (Y_y - X_x)y' - X_y y'^2] + Xy^4 + 4xy^3 Y = 0. \end{aligned} \tag{2.101}$$

Setting the coefficients of y', y'^2, and y'^3 gives

$$Y_{xx} - xy^4(Y_y - 2X_x) + yY_x + Xy^4 + 4xy^3 Y = 0, \tag{2.102a}$$

$$2Y_{xy} - X_{xx} + yX_x + 3xy^4 X_y + Y = 0, \tag{2.102b}$$

$$Y_{yy} - 2X_{xy} + 2yX_y = 0, \tag{2.102c}$$

$$X_{yy} = 0, \tag{2.102d}$$

an over-determined system of equations for the unknowns X and Y. Integrating equation (2.102d) gives

$$X(x, y) = a(x)y + b(x). \tag{2.103}$$

Substituting into (2.102c) gives

$$\begin{aligned} Y_{yy} &= 2X_{xy} - 2yX_y = 2a'(x) - 2a(x)y, \\ Y_y &= -a(x)y^2 + 2a'(x)y + p(x), \\ Y &= -\frac{1}{3}\, a(x)y^3 + a'(x)y^2 + p(x)y + q(x). \end{aligned} \tag{2.104}$$

Substituting (2.103) and (2.104) into equation (2.102b) and regrouping gives

$$\begin{aligned} &3xa(x)y^4 - \frac{1}{3}\, a(x)y^3 + (3a''(x) + b'(x) + p(x))y \\ &\quad + 2p'(x) - b''(x) + q(x) = 0. \end{aligned} \tag{2.105}$$

When we set the coefficients of the various powers of y in (2.105) to zero, we obtain

$$a(x) = 0, \tag{2.106a}$$

$$3a''(x) + b'(x) + p(x) = 0, \tag{2.106b}$$

$$2p'(x) - b''(x) + q(x) = 0. \tag{2.106c}$$

From (2.106a) we see that $a(x) = 0$, which leads to

$$X = b(x), \quad Y = p(x)y + q(x). \tag{2.107}$$

Substituting (2.103) and (2.104) into equation (2.102a) and reorganizing gives

$$(2xb' + b + 3xp)y^4 + 4xqy^3 + p'y^2 + (q' + p'')y + q'' = 0. \tag{2.108}$$

When we set the coefficients of the various powers of y to zero, we get

$$q = 0, \quad p' = 0, \quad 2xb'(x) + b(x) + 3xp = 0. \tag{2.109}$$

The solution of the over-determined system (2.106) and (2.109) is

$$b(x) = cx, \quad p(x) = -c, \quad q(x) = 0, \tag{2.110}$$

where c is constant. This leads to the infinitesimals X and Y as

$$X = cx, \quad Y = -cy. \tag{2.111}$$

Now we find the change of variable that will lead to an equation independent of s. Setting $c = 1$ and solving

$$xr_x - yr_y = 0, \; xs_x - ys_y = 1, \tag{2.112}$$

leads to

$$r = xy, \; s = \ln x,$$

or

$$x = e^s, \; y = re^{-s}. \tag{2.113}$$

In terms of these new variables, equation (2.97) becomes

$$s_{rr} = (r^4 - r^2 + 2r)s_r^3 + (r - 3)s_r^2,$$

a second-order equation independent of s. We note that introducing $q = s_r$ gives

$$q_r = (r^4 - r^2 + 2r)q^3 + (r - 3)q^2, \tag{2.114}$$

a first-order ODE in q.

So far in this section, we have considered two examples. The first was $y'' = 0$ to demonstrate the algorithmic nature of Lie's method. The second ODE was

$$y'' + yy' + xy^4 = 0,$$

where the infinitesimals

$$X = c_1 x, \;\; Y = -c_1 y$$

were obtained. We naturally wonder how sophisticated the infinitesimals X and Y can be. The next example gives us the answer to that question. ■

EXAMPLE 2.19

Consider the following ODE:

$$y'' + 3yy' + y^3 = 0. \tag{2.115}$$

If we denote this equation by Δ, then

$$\Delta = y'' + 3yy' + y^3 = 0,$$

and Lie's invariance condition is

$$\begin{aligned} \Gamma^{(2)}(\Delta)\Big|_{\Delta=0} &= 0, \\ \Gamma^{(2)}(y'' + 3yy' + y^3)\Big|_{y''+3yy'+y^3=0} &= 0, \\ Y_{[xx]} + 3Yy' + 3yY_{[x]} + 3y^2 Y\Big|_{y''+3yy'+y^3=0} &= 0. \end{aligned} \tag{2.116}$$

Substituting the extended infinitesimal transformations

$$\begin{aligned} Y_{[x]} &= D_x(Y) - y' D_x(X) \\ Y_{[xx]} &= D_x(Y_{[x]}) - y'' D_x(X) \end{aligned}$$

into equation (2.116) and expanding gives

$$\begin{aligned} &Y_{xx} + (2Y_{xy} - X_{xx})y' + (Y_{yy} - 2X_{xy})y'^2 - X_{yy}y'^3 \\ &+ (Y_y - 2X_x)y'' - 3X_y y' y'' + 3Yy' \\ &+ 3y(Y_x + (Y_y - X_x)y' - X_y y'^2) + 3y^2 Y = 0. \end{aligned} \tag{2.117}$$

Substituting $y'' = -3yy' - y^3$ into (2.117) gives

$$\begin{aligned} &Y_{xx} + (2Y_{xy} - X_{xx})y' + (Y_{yy} - 2X_{xy})y'^2 - X_{yy}y'^3 \\ &\quad + (Y_y - 2X_x)(-3yy' - y^3) - 3X_y y'(-3yy' - y^3) + 3Yy' \\ &\quad + 3y(Y_x + (Y_y - X_x)y' - X_y y'^2) + 3y^2 Y = 0. \end{aligned} \tag{2.118}$$

Setting the coefficients of y', y'^2, and y'^3 in (2.118) to zero gives

$$Y_{xx} + 2y^3 X_x + 3y^2 Y + 3yY_x - y^3 Y_y = 0, \tag{2.119a}$$

$$2Y_{xy} - X_{xx} + 3yX_x + 3y^3 X_y + 3Y = 0, \tag{2.119b}$$

$$Y_{yy} - 2X_{xy} + 6yX_y = 0, \tag{2.119c}$$

$$X_{yy} = 0, \tag{2.119d}$$

an over-determined system of equations for the unknowns X and Y. Integrating equation (2.119d) gives

$$X(x, y) = a(x)y + b(x). \tag{2.120}$$

Substituting into (2.119c) gives

$$\begin{aligned} Y_{yy} &= 2X_{xy} = 2a'(x) - 6a(x)y, \\ Y_y &= -3a(x)y^2 + 2a'(x)y + p(x), \\ Y &= -a(x)y^3 + a'(x)y^2 + p(x)y + q(x). \end{aligned} \tag{2.121}$$

Substituting (2.120) and (2.121) into equation (2.119b) and regrouping gives

$$3(a''(x) + b'(x) + p(x))y - \left(b''(x) - 2p'(x) - 3q(x)\right).$$

When we set the coefficients of y to zero, we obtain

$$a''(x) + b'(x) + p(x) = 0, \tag{2.122a}$$

$$b''(x) - 2p'(x) - 3q(x) = 0. \tag{2.122b}$$

Substituting (2.120) and (2.121) into equation (2.119a) gives

$$\begin{aligned} &2(a''(x) + b'(x) + p(x))y^3 + (a'''(x) + 3p'(x) + 3q(x))y^2 \\ &\quad + (p''(x) + 3q'(x))y + q''(x) = 0. \end{aligned}$$

Setting the coefficients of y and powers of y to zero gives

$$a''(x) + b'(x) + p(x) = 0, \tag{2.123a}$$

$$a'''(x) + 3p'(x) + 3q(x) = 0, \tag{2.123b}$$

$$p''(x) + 3q'(x) = 0, \tag{2.123c}$$

$$q''(x) = 0. \tag{2.123d}$$

The solution of (2.123d) is given simply as

$$q(x) = q_1 x + q_2. \tag{2.124}$$

When we substitute (2.124) into equation (2.123c) and integrate, we obtain

$$p(x) = -\frac{3}{2} q_1 x^2 + p_1 x + p_2. \tag{2.125}$$

Substituting (2.124) and (2.125) into equation (2.122b) leads to

$$b''(x) = -3q_1 x + 2p_1 + 3q_2,$$

which, upon integration, gives

$$b(x) = -\frac{1}{2} q_1 x^3 + \frac{1}{2}(2p_1 + 3q_2)x^2 + b_1 x + b_2. \tag{2.126}$$

This only leaves equations (2.122a), (2.123a), and (2.123b), and the first two are identical. Equation (2.122a) gives

$$a''(x) = q_1 x^2 - 3(p_1 + q_2)x - (b_1 + p_2),$$

which, upon integration, gives

$$a(x) = \frac{1}{4} q_1 x^4 - \frac{1}{2}(p_1 + q_2)x^3 - \frac{1}{2}(b_1 + p_2)x^2 + a_1 x + a_2. \tag{2.127}$$

The remaining equation (2.123b) is shown to be automatically satisfied. With a, b, p, and q given in equations (2.124)–(2.127), the infinitesimals X and Y are obtained through (2.120) and (2.121). For simplicity, we will reset the constants as follows:

$$a_1 = c_2,\ a_2 = c_1,\ b_1 = c_7,\ b_2 = c_6,$$

$$p_1 = -6c_4 - 2c_8,\ p_2 = -c_7 - 2c_3,$$

$$q_1 = 4c_5,\ q_2 = 4c_4 + 2c_8.$$

This gives the following eight-parameter family of infinitesimals:

$$\begin{aligned}
X &= (c_1 + c_2 x + c_3 x^2 + c_4 x^3 + c_5 x^4) y \\
&\quad + c_6 + c_7 x + c_8 x^2 - 2c_5 x^3, \\
Y &= -(c_1 + c_2 x + c_3 x^2 + c_4 x^3 + c_5 x^4) y^3 \\
&\quad + (c_2 + 2c_3 x + 3c_4 x^2 + 4c_5 x^3) y^2 \\
&\quad - (2c_3 + c_7 + (6c_4 + 2c_8) x + 6c_5 x^2) y \\
&\quad + 4c_4 + 2c_8 + 4c_5 x,
\end{aligned}$$

where $c_1 - c_8$ are arbitrary constants. For several constants, we will construct associated infinitesimal operators and for each a change of variables will be obtained. These infinitesimal operators will be denoted by Γ_i, where the coefficients of the operator are obtained by setting $c_i = 1$ and all other constants to zero in X and Y. This gives

$$\begin{aligned}
\Gamma_1 &= y\frac{\partial}{\partial x} - y^3\frac{\partial}{\partial y}, \\
\Gamma_2 &= xy\frac{\partial}{\partial x} + (y^2 - xy^3)\frac{\partial}{\partial y}, \\
\Gamma_3 &= x^2 y\frac{\partial}{\partial x} + (-x^2 y^3 + 2xy^2 - 2y)\frac{\partial}{\partial y}, \\
\Gamma_4 &= x^3 y\frac{\partial}{\partial x} + (-x^3 y^3 + 3x^2 y^2 - 6xy + 4)\frac{\partial}{\partial y}, \\
\Gamma_5 &= (x^4 y - 2x^3)\frac{\partial}{\partial x} + (-x^4 y^3 + 4x^3 y^2 - 6x^2 y + 4x)\frac{\partial}{\partial y}, \\
\Gamma_6 &= \frac{\partial}{\partial x}, \\
\Gamma_7 &= x\frac{\partial}{\partial x} - y\frac{\partial}{\partial y}, \\
\Gamma_8 &= x^2\frac{\partial}{\partial x} + (2 - 2xy)\frac{\partial}{\partial y}.
\end{aligned}$$

We now find a change of variables for some of the infinitesimal operators shown earlier. These are constructed by solving (2.34) or in terms of the infinitesimal operator Γ

$$\Gamma r = 0, \quad \Gamma s = 1. \tag{2.128}$$

In particular, we will consider Γ_1, Γ_2, Γ_6, and Γ_7, as these are probably the easiest to solve

$$\Gamma_1 = y\frac{\partial}{\partial x} - y^3\frac{\partial}{\partial y}.$$

In this case, we need to solve

$$\begin{aligned} yr_x - y^3 r_y &= 0, \\ ys_x - y^3 s_y &= 1. \end{aligned}$$

This leads to

$$r = x - \frac{1}{y}, \quad s = \frac{1}{2y^2},$$

or

$$x = r + \sqrt{2s}, \quad y = \frac{1}{\sqrt{2s}}. \tag{2.129}$$

Under the change of variables (2.129), equation (2.115) becomes

$$s_{rr} = 1,$$

a second-order equation independent of s

$$\Gamma_2 = xy\frac{\partial}{\partial x} + (y^2 - xy^3)\frac{\partial}{\partial y}.$$

In this case, we need to solve

$$\begin{aligned} xyr_x + (y^2 - xy^3)r_y &= 0, \\ xys_x + (y^2 - xy^3)s_y &= 1. \end{aligned}$$

This leads to

$$r = \frac{x}{y} - \frac{1}{2}x^2, \quad s = x - \frac{1}{y},$$

or

$$x = s + \sqrt{2r + s^2}, \quad y = \frac{1}{\sqrt{2r + s^2}}. \tag{2.130}$$

Under the change of variables (2.130), equation (2.115) becomes

$$s_{rr} = 0,$$

which is trivially solved.

$$\Gamma_6 = \frac{\partial}{\partial x}$$

In this case, we need to solve

$$r_x = 0, \quad s_x = 1.$$

This leads to

$$r = y, \quad s = x,$$

or

$$x = s, \quad y = r. \tag{2.131}$$

Under the change of variables (2.131), equation (2.115) becomes

$$s_{rr} = r^3 s_r^3 + 3rs_r^2,$$

a second-order equation independent of s. Therefore, introducing $q = s_r$ gives

$$q_r = r^3 q^3 + 3rq^2,$$

a first-order ODE in q.

$\boldsymbol{\Gamma_7 = x\frac{\partial}{\partial x} - y\frac{\partial}{\partial y}}$.

In this case, we need to solve

$$xr_x - yr_y = 0, \quad xs_x - ys_y = 1.$$

This leads to

$$r = xy, \quad s = \ln x,$$

or

$$x = e^s, \quad y = re^{-s}. \tag{2.132}$$

Under the change of variables (2.132), equation (2.115) becomes

$$s_{rr} = (r-1)(r^2 - 2r)s_r^3 + 3(r-1)s_r^2,$$

a second order equation independent of s. Therefore, introducing $q = s_r$ gives

$$q_r = (r-1)(r^2 - 2r)q^3 + 3(r-1)q^2,$$

a first-order ODE in q. ■

EXERCISES

1. Find constants a and b such that

$$\frac{d^2y}{dx^2} + \frac{n}{x}\frac{dy}{dx} + y\frac{dy}{dx} = 0, \quad n = 0, 1, \text{ and } 2,$$

is invariant under the Lie group of transformations

$$\bar{x} = e^{a\varepsilon}x, \quad \bar{y} = e^{b\varepsilon}y.$$

Calculate the infinitesimal transformations associated with this Lie group. Find the canonical coordinates that will reduce this ODE to one that is independent of s and then further reduce the order.

2. Calculate the eight-parameter family of infinitesimals, leaving

$$y^2 y'' + 2xy'^3 = 0$$

invariant. Choose any three and show that they lead to new variables that reduce the original equation to one that is of lower dimension. Solve any of the equations.

3. Find the infinitesimals X and Y, leaving the following ODEs invariant

$$(i)\ \frac{d^2y}{dx^2} = \frac{y^n}{x^2}, \quad n \in \mathbb{R}/\{0, 1\}$$

$$(ii)\ \frac{d^2y}{dx^2} + \frac{n}{x}\frac{dy}{dx} = e^y.$$

For each, reduce to first order. (Be careful, there might be special cases for n.)

4*. Determine the symmetries of the linear ODE

$$\frac{d^2y}{dx^2} + f(x)y = 0.$$

5*. Determine the forms of $F(x)$ and the associated infinitesimals X and Y, leaving the following ODE invariant

$$\frac{d^2y}{dx^2} = \frac{F(x)}{y^2}.$$

Note: If you get an equation of the form

$$\frac{F'}{F} = \frac{c_4 x + c_5}{c_1 x^2 + c_2 x + c_3},$$

then there are several cases that must be considered: $c_1 = 0$ and $c_1 \neq 0$. Within the latter, does the quadratic equation in x have real distinct roots, repeated roots or complex roots? These cases will give different forms of F.

6*. Classify the symmetries of the ODE

$$2\frac{\mathrm{d}}{\mathrm{d}x}\left(K(y)\frac{\mathrm{d}y}{\mathrm{d}x}\right) + x\frac{\mathrm{d}y}{\mathrm{d}x} = 0,$$

according to the form of $K(y)$. This ODE arises from a seeking invariance of a nonlinear heat equation (Bluman and Reid [3]).

2.6 HIGHER ORDER EQUATIONS

As we become more comfortable with Lie's method of symmetry analysis, it is natural to extend the method to higher order equations. After a few examples, one gets the idea that computer algebra systems could be a valuable tool in manipulating and solving the determining equations (those equations that define the infinitesimals X and Y). This becomes clear with the following example.

EXAMPLE 2.20

Consider

$$y''' + yy'' = 0. \tag{2.133}$$

This ODE is known as Blasius's equation and appears in the study of steady $2-D$ boundary layers in fluid mechanics. If we denote (2.133) by Δ, then

$$\Delta = y''' + yy'' = 0.$$

Lie's invariance condition is

$$\begin{aligned} \Gamma^{(3)}(\Delta)\Big|_{\Delta=0} &= 0, \\ \Gamma^{(3)}(y''' + yy'')\Big|_{y'''+yy''=0} &= 0, \\ Y_{[xxx]} + Yy'' + yY[xx]\Big|_{y'''+yy''=0} &= 0. \end{aligned} \tag{2.134}$$

Substituting the extended infinitesimal transformations

$$Y_{[x]} = D_x(Y) - y' D_x(X)$$
$$Y_{[xx]} = D_x(Y_{[x]}) - y'' D_x(X)$$
$$Y_{[xxx]} = D_x(Y_{[xx]}) - y''' D_x(X)$$

into equation (2.134) and expanding gives

$$\begin{aligned} &Y_{xxx} + (3Y_{xxy} - X_{xxx})y' + 3(Y_{xyy} - X_{xxy})y'^2 \\ &+ (Y_{yyy} - 3X_{xyy})y'^3 - X_{yyy}y'^4 + 3(Y_{xy} - X_{xx})y'' \\ &+ 3(Y_{yy} - 3X_{xy})y'y'' - 6X_{yy}y'^2y'' - 3X_y y''^2 \\ &+ (Y_y - 3X_x)y''' - 4X_y y' y''' \\ &+ yY_{xx} + y(2Y_{xy} - X_{xx})y' + y(Y_{yy} - 2X_{xy})y'^2 - yX_{yy}y'^3 \\ &+ y(Y_y - 2X_x)y'' - 3yX_y y' y'' + Yy'' = 0. \end{aligned} \tag{2.135}$$

Substituting $y''' = -yy''$ into (2.135) gives

$$\begin{aligned} &Y_{xxx} + (3Y_{xxy} - X_{xxx})y' + 3(Y_{xyy} - X_{xxy})y'^2 \\ &+ (Y_{yyy} - 3X_{xyy})y'^3 - X_{yyy}y'^4 + 3(Y_{xy} - X_{xx})y'' \\ &+ 3(Y_{yy} - 3X_{xy})y'y'' - 6X_{yy}y'^2y'' - 3X_y y''^2 \\ &- (Y_y - 3X_x)yy'' + 4X_y yy' y'' \\ &+ yY_{xx} + y(2Y_{xy} - X_{xx})y' + y(Y_{yy} - 2X_{xy})y'^2 - yX_{yy}y'^3 \\ &+ y(Y_y - 2X_x)y'' - 3yX_y y' y'' + Yy'' = 0. \end{aligned} \tag{2.136}$$

note the appearance of derivatives y' and y'' in (2.136).

When we set the coefficients of y', y'^2, y'^3, y'^4, y'', $y'y''$, y'^2y'', and y''^2 to zero, it gives

$$3Y_{xxy} - X_{xxx} + y(2Y_{xy} - X_{xx}) = 0, \tag{2.137a}$$

$$3(Y_{xyy} - X_{xxy}) + y(Y_{yy} - 2X_{xy}) = 0, \tag{2.137b}$$

$$Y_{yyy} - 3X_{xyy} - yX_{yy} = 0, \tag{2.137c}$$

$$-X_{yyy} = 0, \tag{2.137d}$$

$$3(Y_{xy} - X_{xx}) + yX_x + Y = 0, \tag{2.137e}$$

$$3(Y_{yy} - 3X_{xy}) + yX_y = 0, \tag{2.137f}$$

$$-6X_{yy} = 0, \tag{2.137g}$$

$$-3X_y = 0, \tag{2.137h}$$

$$Y_{xxx} + yY_{xx} = 0. \tag{2.137i}$$

This is an over-determined system of equations for the unknowns X and Y. Equation (2.137h) gives

$$X(x, y) = a(x), \tag{2.138}$$

where a is arbitrary. From (2.137h) and (2.137f), we see that

$$Y_{yy} = 0, \tag{2.139}$$

from which we obtain

$$Y(x, y) = b(x)y + c(x), \tag{2.140}$$

where b and c are arbitrary. Substituting (2.138) and (2.140) into equation (2.137e) gives

$$(a' + b)y - 3a'' + 3b' + c = 0,$$

from which we obtain

$$a' + b = 0, \tag{2.141a}$$

$$-3a'' + 3b' + c = 0. \tag{2.141b}$$

From equation (2.137a), we get

$$(2b' - a'')y + 3b'' - a''' = 0, \tag{2.142}$$

from which we obtain

$$2b' - a'' = 0, \tag{2.143a}$$

$$3b'' - a''' = 0. \tag{2.143b}$$

From (2.141) and (2.143), we deduce

$$a'' = 0, \quad b' = 0,$$

from which we obtain

$$a(x) = c_1 x + c_2, \quad b(x) = -c_1,$$

where c_1 and c_2 are arbitrary constants. From equation (2.141b)

$$c(x) = 0,$$

thus giving the infinitesimals X and Y as

$$X(x, y) = c_1 x + c_2, \tag{2.144a}$$

$$Y(x, y) = -c_1 y. \tag{2.144b}$$

Substitution of (2.144a) into the remaining equations in (2.137) shows that they are automatically satisfied. The infinitesimal operator for each constant is

$$\Gamma_1 = x\frac{\partial}{\partial x} - y\frac{\partial}{\partial y}, \quad \Gamma_2 = \frac{\partial}{\partial x}.$$

Now we find new variables r and s for each infinitesimal operator shown earlier. These are constructed by solving

$$\Gamma r = 0, \ \Gamma s = 1.$$

We will consider each separately.

$\boldsymbol{\Gamma_1 = x\frac{\partial}{\partial x} - y\frac{\partial}{\partial y}}$.

In this case, we need to solve

$$xr_x - yr_y = 0, \quad xs_x - ys_y = 1.$$

This leads to

$$r = xy, \quad s = \ln x$$

or

$$x = \mathrm{e}^s, \quad y = r\mathrm{e}^{-s}.$$

In terms of these new coordinates, equation (2.133) becomes

$$s_r s_{rrr} + rs_r^2 s_{rr} - 3s_{rr}^2 = 0, \tag{2.145}$$

a second-order equation independent of s. Setting $q = s_r$ in (2.145) gives

$$qq_{rr} + rq^2 q_r - 3q_r^2 = 0. \tag{2.146}$$

We note that equation (2.146) admits the Lie Group

$$\bar{r} = \mathrm{e}^{a\varepsilon} r, \quad \bar{q} = \mathrm{e}^{-2a\varepsilon} q,$$

and hence the infinitesimal operator

$$\Gamma = r\frac{\partial}{\partial r} - 2q\frac{\partial}{\partial q}.$$

Thus, we can reduce (2.146) further. Solving

$$rt_x - 2qt_y = 0, \quad ru_x - 2qu_y = 1,$$

leads to

$$r = \mathrm{e}^u, \quad t - r^2 q$$

or

$$r = \mathrm{e}^u, \quad q = t\mathrm{e}^{-2u}.$$

In terms of these new variables, equation (2.146) becomes

$$tu_{tt} + (2t^3 + 6t^2)u_t^3 - (t^2 + 7t)u_t^2 + 3u_t = 0,$$

which is reduced to a first-order equation if $v = u_t$.

$\mathbf{\Gamma_2 = \frac{\partial}{\partial x}}$.

In this case, we need to solve

$$r_x = 0, \quad s_x = 1.$$

This leads to

$$r = y, \quad s = x,$$

or

$$x = s, \quad y = r.$$

In terms of these new variables, equation (2.133) becomes

$$s_r s_{rrr} - 3s_{rr}^2 + (r - 6)s_r^2 s_{rr} + (3r - 11)s_r^4 - (2r^2 - 6r)s_r^5 = 0,$$

which, under the substitution $v = s_r$, gives an equation that is second order but still quite difficult to solve in general. ■

EXERCISES

1. Find the infinitesimals X and Y, leaving the following ODEs invariant

$$(i)\ y''' + yy'' - y'^2 = 0,$$

$$(ii)\ y''' + 4yy'' + 3y'^2 + 6y^2y' + y^4 = 0.$$

$$(iii)\ y''' = y^{-3}.$$

For each, reduce to one that is second order.

2. Calculate the symmetries of the Chazy equation (see Clarkson and Olver [4])

$$y''' = 2yy'' - 3y'^2.$$

3. Calculate the symmetries of the equation

$$\left(yy'\left(\frac{y}{y'}\right)''\right)' = 0.$$

which arises from the symmetries of the wave equation (see Bluman and Kumei [5]).

2.7 ODE SYSTEMS

In this section, methods for the solution of ODEs using Lie symmetries are extended to systems of ODEs.

2.7.1 First Order Systems

Consider

$$\dot{x} = f(t, x, y), \tag{2.147a}$$

$$\dot{y} = g(t, x, y), \tag{2.147b}$$

where an overdot denotes $\frac{\mathrm{d}(\,)}{\mathrm{d}t}$. We consider invariance of this system under the infinitesimal transformations

$$\bar{t} = t + \varepsilon T(t, x, y) + O(\varepsilon^2), \tag{2.148a}$$

$$\bar{x} = x + \varepsilon X(t, x, y) + O(\varepsilon^2), \tag{2.148b}$$

$$\bar{y} = y + \varepsilon Y(t, x, y) + O(\varepsilon^2). \tag{2.148c}$$

By extension of our previous work on single equations, the invariance of the system (2.147) under the transformations (2.148) leads to Lie's invariance condition

$$\Gamma^{(1)}\Delta\Big|_{\Delta=0} = 0,$$

where Δ refers to the system (2.147). For systems like (2.147), the infinitesimal operator Γ is defined as

$$\Gamma = T\frac{\partial}{\partial t} + X\frac{\partial}{\partial x} + Y\frac{\partial}{\partial y} \tag{2.149}$$

with the first extension defined as

$$\Gamma^{(1)} = \Gamma + X_{[t]}\frac{\partial}{\partial \dot{x}} + Y_{[t]}\frac{\partial}{\partial \dot{y}}. \tag{2.150}$$

Higher order extensions are defined similarly. For example, the second extension is

$$\Gamma^{(2)} = \Gamma^{(1)} + X_{[tt]}\frac{\partial}{\partial \ddot{x}} + Y_{[tt]}\frac{\partial}{\partial \ddot{y}}.$$

In (2.150), the extended infinitesimal transformations are

$$X_{[t]} = D_t(X) - \dot{x}D_t(T), \tag{2.151a}$$

$$Y_{[t]} = D_t(Y) - \dot{y}D_t(T), \tag{2.151b}$$

where the total differential operator D_t is defined as

$$D_t = \frac{\partial}{\partial t} + \dot{x}\frac{\partial}{\partial x} + \dot{y}\frac{\partial}{\partial y} + \ddot{x}\frac{\partial}{\partial \dot{x}} + \ddot{y}\frac{\partial}{\partial \dot{y}} + \cdots.$$

The second extended infinitesimal transformations are

$$X_{[tt]} = D_t\left(X_{[t]}\right) - \ddot{x}D_t(T),$$

$$Y_{[tt]} = D_t\left(Y_{[t]}\right) - \ddot{y}D_t(T).$$

Once the infinitesimals T, X, and Y have been found, we must solve

$$Tr_t + Xr_x + Yr_y = 0,$$
$$Tu_t + Xu_x + Yu_y = 0,$$
$$Tv_t + Xv_x + Yv_y = 1.$$

This will then give a set of new variables that will transform the given system to one that is independent of v.

EXAMPLE 2.21

Consider the system

$$\dot{x} = 2xy, \tag{2.152a}$$
$$\dot{y} = x^2 + y^2. \tag{2.152b}$$

Examination of this system shows that it could easily be rewritten as

$$\frac{\mathrm{d}y}{\mathrm{d}x} = \frac{x^2 + y^2}{2xy},$$

which is both homogeneous and of Bernoulli type, and so can be solved by standard methods. Symmetry methods are usually used for equations which are not amenable to the standard methods. Thus, the system of equations (2.152a) is solved here using symmetry methods in order to demonstrate Lie's method.
Applying (2.149) and (2.150) to the system (2.152a) yields

$$X_{[t]} = 2Xy + 2xY \tag{2.153a}$$
$$Y_{[t]} = 2xX + 2yY, \tag{2.153b}$$

and substituting the extended transformations (2.151) yields

$$X_t + \dot{x}X_x + \dot{y}Y_y - \dot{x}(T_t + \dot{x}T_x + \dot{y}T_y) = 2yX + 2xY \tag{2.154a}$$
$$Y_t + \dot{x}Y_x + \dot{y}Y_y - \dot{y}(T_t + \dot{x}T_x + \dot{y}T_y) = 2xX + 2yY. \tag{2.154b}$$

Substituting (2.152) into (2.154) finally yields

$$X_t + (X_x - T_t)2xy + (x^2 + y^2)X_y - (2xy)^2 T_x - 2xy(x^2 + y^2)T_y = 2Xy + 2xY \tag{2.155a}$$

$$Y_t + 2xyY_x + (x^2+y^2)(Y_y - T_t) - 2xy(x^2+y^2)T_x + (x^2+y^2)^2 T_y = 2xX + 2yY. \quad (2.155b)$$

As the system (2.155) is difficult to solve in general, we seek special solutions by assuming that

$$T = T(t), \quad X = X(x), \quad Y = Y(y). \quad (2.156)$$

We note that these are chosen only to reduce the complexity of (2.155) and other choices could be made. Under the assumptions in (2.156), equation (2.155) becomes

$$2xy\left(X_x - T_t\right) = 2yX + 2xY, \quad (2.157a)$$

$$(x^2+y^2)(Y_y - T_t) = 2xX + 2yY, \quad (2.157b)$$

which is obviously much simpler. We now need some nontrivial infinitesimals (those that are not identically zero). If we take the partial derivative of (2.157a) with respect to t, we find

$$-2xyT_{tt} = 0 \quad \Rightarrow \quad T_{tt} = 0.$$

Thus,

$$T(t) = at + b, \quad (2.158)$$

where a and b are constant. Inserting (2.158) into (2.157), we find

$$2xy\left(X_x - a\right) = 2yX + 2xY, \quad (2.159a)$$

$$(x^2+y^2)(Y_y - a) = 2xX + 2yY. \quad (2.159b)$$

Partially differentiating (2.159a) with respect to y twice yields

$$Y''(y) = 0.$$

Thus,

$$Y(y) = py + q, \quad (2.160)$$

where p and q are constant.

In the process of annihilating (2.159a) by differentiating, it is possible that we introduced additional information into the solution. Therefore, we insert (2.160) into our starting point (2.159a), yielding

$$2xy(X_x - a) = 2yX + 2x(py+q),$$

or

$$(2xX_x - 2ax - 2X - 2px)y - 2xq = 0. \tag{2.161}$$

As (2.161) must be satisfied for all values of y, we immediately see that

$$xX_x - ax - X - px = 0, \tag{2.162a}$$

$$q = 0. \tag{2.162b}$$

Using $q = 0$, we find from (2.160) that $Y = py$. In other words, our simplifications introduced an additional constant that should not have been there. Inserting this result into (2.159b) yields

$$(p - a)(x^2 + y^2) = 2xX + 2py^2.$$

Grouping like terms in powers of y gives

$$(-p - a)y^2 + (p - a)x^2 - 2xX = 0.$$

As this equation must be satisfied for all values of y and the fact that X is independent of y, then

$$p + a = 0,$$

$$(p - a)x^2 - 2xX = 0,$$

which leads to $p = -a$ and $X = ax$, noting that (2.162a) is automatically satisfied. Therefore, the infinitesimals are in the given equation:

$$T = at + b, \quad X = -ax, \quad Y = -ay. \tag{2.163}$$

Of course, other infinitesimals could be found. In fact, there is an infinite set of infinitesimals.

Our next task is to find a change of variables. For convenience, we set $a = 1$ and $b = 0$ in (2.163) and solve

$$tr_t - xr_x - yr_y = 0,$$

$$tu_t - xu_x - yu_y = 0,$$

$$tv_t - xv_x - yv_y = 1.$$

The solution of these linear partial differential equations is

$$r = R(tx, ty), \quad u = U(tx, ty), \quad v = \ln t + V(tx, ty).$$

As we have some flexibility in our choice of R, U, and V, we consider two different choices.

Choice 1 If we choose

$$r = tx, \quad u = ty, \quad v = \ln t,$$

then

$$t = e^{v}, \quad y = \frac{u}{t} = ue^{-v}, \quad x = \frac{r}{t} = re^{-v},$$

and (2.152) becomes

$$u_r = \frac{u^2 + u + 2r^2}{(2u + 1)r}, \quad v_r = \frac{1}{r(1 + 2u)}.$$

We see the system decouples; however, we need the solution for u to find the solution for v. A different choice of variables could lead to an even simpler set of equations to solve as the next example illustrates.

Choice 2 If we choose

$$r = \frac{y}{x}, \quad u = tx, \quad v = -\ln x, \tag{2.164}$$

then

$$t = ue^{v}, \quad x = e^{-v}, \quad y = re{-}v,$$

and (2.152) becomes

$$u_r = -\frac{1 + 2ru}{r^2 - 1}, \quad v_r = \frac{2r}{r^2 - 1}.$$

The solution of each is easily obtained giving

$$u = \frac{-r + c_1}{r^2 - 1}, \quad v = \ln\left|r^2 - 1\right| + c_2. \tag{2.165}$$

Now all that remains is to transform back to our original variables (2.164). Substituting (2.164) into (2.165) gives

$$tx = \frac{-\dfrac{y}{x} + c_1}{\left(\dfrac{y}{x}\right)^2 - 1}, \quad \ln x = \ln\left|\frac{y^2}{x^2} - 1\right| + c_2, \tag{2.166}$$

and solving (2.166) for x and y gives

$$x = \frac{a}{(at + b)^2 - 1} \text{ and } y = \frac{-a(at + b)}{(at + b)^2 - 1},$$

where we have set $c_1 = -b$ and $c_2 = -\ln a$. ■

2.7.2 Higher Order Systems

In this section, we consider the symmetries of higher order systems of ordinary differential equations. In particular, we will consider second-order systems but the analysis is not restricted to only this. Consider

$$\ddot{x} = P(x, y), \quad P_y \neq 0, \tag{2.167a}$$

$$\ddot{y} = Q(x, y), \quad Q_x \neq 0. \tag{2.167b}$$

Note that in this example, P and Q are independent of t, $\dot{x}$, and $\dot{y}$. In general, these terms could be included. Invariance of (2.167) is conveniently written as

$$\Gamma^{(2)}\Delta\Big|_{\Delta=0} = 0, \tag{2.168}$$

where Δ is the system (2.167). The operator Γ is defined as before (see (2.149))

$$\Gamma = T\frac{\partial}{\partial t} + X\frac{\partial}{\partial x} + Y\frac{\partial}{\partial y}, \tag{2.169}$$

and $\Gamma^{(1)}$ and $\Gamma^{(2)}$ are extensions to the operator Γ in (2.169), namely

$$\Gamma^{(1)} = \Gamma + X_{[t]}\frac{\partial}{\partial \dot{x}} + Y_{[t]}\frac{\partial}{\partial \dot{y}},$$

$$\Gamma^{(2)} = \Gamma^{(1)} + X_{[tt]}\frac{\partial}{\partial \ddot{x}} + Y_{[tt]}\frac{\partial}{\partial \ddot{y}}.$$

The invariance condition (2.168) is given by

$$X_{[tt]} = P_x X + P_y Y, \tag{2.170a}$$

$$Y_{[tt]} = Q_x X + Q_y Y, \tag{2.170b}$$

where the extended transformations are given by

$$X_{[tt]} = D_t(X_{[t]}) - \ddot{x}D_t(T), \tag{2.171a}$$

$$Y_{[tt]} = D_t(Y_{[t]}) - \ddot{y}D_t(T), \tag{2.171b}$$

and the total differential operator D_t is given by

$$D_t = \frac{\partial}{\partial t} + \dot{x}\frac{\partial}{\partial x} + \dot{y}\frac{\partial}{\partial y} + \ddot{x}\frac{\partial}{\partial \dot{x}} + \ddot{y}\frac{\partial}{\partial \dot{y}} + \ldots.$$

Substituting (2.171) into (2.170) and expanding gives

$$\begin{aligned} &X_{tt} + \left(2X_{tx} - T_{tt}\right)\dot{x} + 2X_{ty}\dot{y} + \left(X_{xx} - 2T_{tx}\right)\dot{x}^2 \\ &\quad + 2\left(X_{xy} - T_{ty}\right)\dot{x}\dot{y} + X_{yy}\dot{y}^2 - T_{xx}\dot{x}^3 - 2T_{xy}\dot{x}^2\dot{y} \\ &\quad - T_{yy}\dot{x}\dot{y}^2 + \left(X_x - 2T_t\right)\ddot{x} + X_y\ddot{y} - 3T_x\dot{x}\ddot{x} \\ &\quad - 2T_y\dot{y}\ddot{x} - T_y\dot{x}\ddot{y} = XP_x + YP_y \end{aligned} \qquad (2.172)$$

and

$$\begin{aligned} &Y_{tt} + \quad +2Y_{ty}\dot{x} + \left(2Y_{ty} - T_{tt}\right)\dot{y} + Y_{xx}\dot{x}^2 \\ &\quad + 2\left(Y_{xy} - T_{tx}\right)\dot{x}\dot{y} + \left(Y_{yy} - 2T_{ty}\right)\dot{y}^2 + T_{xx}\dot{x}^2\dot{y} \\ &\quad - 2T_{xy}\dot{x}\dot{y}^2 - T_{yy}\dot{y}^3 + Y_x\ddot{x} + \left(Y_y - 2T_t\right)\ddot{y} \\ &\quad - 2T_x\dot{x}\ddot{y} - T_x\dot{y}\ddot{x} - 3T_y\dot{y}\ddot{y} = XQ_x + YQ_y. \end{aligned} \qquad (2.173)$$

Substituting $\ddot{x} = P(x, y)$ and $\ddot{y} = Q(x, y)$ into (2.172) and (2.173) gives two expressions involving multinomials in $\dot{x}$ and $\dot{y}$, namely $\dot{x}$, $\dot{y}$, $\dot{x}^2$, $\dot{x}\dot{y}$, $\dot{y}^2$, $\dot{x}^3$, etc. Setting the coefficient of each to zero gives the following set of determining equations for T, X and Y:

$$X_{tt} + \left(X_x - 2T_t\right)P + X_yQ = XP_x + YP_y, \qquad (2.174a)$$

$$2X_{tx} - T_{tt} - 3T_xP - T_yQ = 0, \qquad (2.174b)$$

$$X_{ty} - T_yP = 0, \qquad (2.174c)$$

$$X_{xx} - 2T_{tx} = 0, \qquad (2.174d)$$

$$X_{xy} - T_{ty} = 0, \qquad (2.174e)$$

$$X_{yy} = 0, \qquad (2.174f)$$

$$T_{xx} = 0, \qquad (2.174g)$$

$$T_{xy} = 0, \qquad (2.174h)$$

$$T_{yy} = 0, \qquad (2.174i)$$

and

$$Y_{tt} + Y_x P + \left(Y_y - 2T_t\right) Q = XQ_x + YQ_y, \qquad (2.175a)$$

$$2Y_{ty} - T_{tt} - T_x P - 3T_y Q = 0, \qquad (2.175b)$$

$$Y_{tx} - T_x Q = 0, \qquad (2.175c)$$

$$Y_{xx} = 0, \qquad (2.175d)$$

$$Y_{xy} - T_{tx} = 0, \qquad (2.175e)$$

$$Y_{yy} - 2T_{ty} = 0, \qquad (2.175f)$$

$$T_{xx} = 0, \qquad (2.175g)$$

$$T_{xy} = 0, \qquad (2.175h)$$

$$T_{yy} = 0. \qquad (2.175i)$$

Note that the last three equations of each set are identical. Before we consider a particular example involving a particular P and Q, we find that we can actually perform quite a bit of analysis on the system of equations (2.174) and (2.175).

Differentiating (2.174c) with respect to y gives

$$X_{tyy} - T_{yy} P - T_y P_y = 0, \qquad (2.176)$$

while differentiating (2.175c) with respect to x gives

$$Y_{txx} - T_{xx} Q - T_x Q_x = 0. \qquad (2.177)$$

By virtue of (2.174f), (2.175d), and the last three equations of either (2.174) or (2.175), we arrive at

$$T_x = 0, \quad T_y = 0, \quad \Rightarrow \quad T = T(t). \qquad (2.178)$$

This results in the fact that the equations (2.174c)–(2.174i) and (2.175c)–(2.175i) are easy to simplify and solve. Thus,

$$X = a(t)x + by + c(t), \qquad (2.179a)$$

$$Y = px + q(t)y + r(t). \qquad (2.179b)$$

Note that the constants b and p arise due to (2.174c) and (2.175c). From (2.174b), (2.175b) gives

$$2a'(t) = T''(t), \quad 2q'(t) = T''(t), \tag{2.180}$$

which, upon integration, gives

$$a(t) = \frac{1}{2}T'(t) + c_1, \quad q(t) = \frac{1}{2}T'(t) + c_2, \tag{2.181}$$

where c_1 and c_2 are arbitrary constants. Thus,

$$X = \left(\frac{1}{2}T'(t) + c_1\right)x + by + c(t), \tag{2.182a}$$

$$Y = px + \left(\frac{1}{2}T'(t) + c_2\right)y + r(t). \tag{2.182b}$$

We now have only equations (2.174a) and (2.175a) left which explicitly involve P and Q. At this point, we consider a particular example.

EXAMPLE 2.22

Consider the following system

$$\ddot{x} = \frac{x}{(x^2 + y^2)^2}, \quad \ddot{y} = \frac{y}{(x^2 + y^2)^2}. \tag{2.183}$$

We identify P and Q in this system as

$$P = \frac{x}{(x^2 + y^2)^2}, \quad Q = \frac{y}{(x^2 + y^2)^2}.$$

From (2.174a) and (2.175a), we set the coefficients of multinomials involving x and y to zero and obtain

$$c_1 = 0, \quad c_2 = 0, \quad b + p = 0, \quad c(t) = 0, \quad r(t) = 0, \quad T'''(t) = 0, \tag{2.184}$$

from which we obtain

$$T = k_2 t^2 + 2k_1 t + k_0, \tag{2.185}$$

where k_0, k_1, and k_2 are arbitrary constants. From (2.182), (2.184), and (2.185), we find the infinitesimals

$$\begin{aligned} T &= k_2 t^2 + 2k_1 t + k_0, \\ X &= \left(k_2 t + k_1\right)x + k_3 y, \\ Y &= -k_3 x + \left(k_2 t + k_1\right)y. \end{aligned} \tag{2.186}$$

For a particular example, we consider the case when $k_0 = 0, k_1 = 0, k_2 = 0$, and $k_3 = 1$. We introduce new coordinates r, u, and v such that

$$yr_x - xr_y = 0, \quad yu_x - xu_y = 0, \quad yv_x - xv_y = 1. \tag{2.187}$$

Solving (2.187) gives

$$r = R\left(t, x^2 + y^2\right), \quad u = U\left(t, x^2 + y^2\right), \quad v = \tan^{-1}\frac{y}{x} + V\left(t, x^2 + y^2\right). \tag{2.188}$$

where R, U and V are arbitrary. Choosing these arbitrary functions such that

$$r = t, \quad u = \sqrt{x^2 + y^2}, \quad v = \tan^{-1}\frac{y}{x} \tag{2.189}$$

transforms the system (2.183) to the new system

$$u_{rr} = \frac{u^4 v_r^2 + 1}{u^3}, \quad v_{rr} = -\frac{2u_r v_r}{u}, \tag{2.190}$$

a system independent of the variable v. ■

EXERCISES

1. Calculate the scaling symmetries for the following system of ODEs arising in modeling the formation of polymers (see Sediawan and Megawati [6])

$$\begin{aligned}
\dot{x} &= -2x^2 - xy - xz, \\
\dot{y} &= x^2 - xy - 2y^2 - yz, \\
\dot{z} &= xy - xz - yz - 2z^2.
\end{aligned}$$

2. Calculate the symmetries for the following system of ODEs:

$$\begin{aligned}
\ddot{x} &= \dot{x}\sqrt{\dot{x}^2 + \dot{y}^2}, \\
\ddot{y} &= \dot{y}\sqrt{\dot{x}^2 + \dot{y}^2}.
\end{aligned}$$

3. Calculate the symmetries for the following system of ODEs arising from a reduction of the cubic Schrodinger equation

$$\begin{aligned}
\ddot{x} - \dot{y} + x(x^2 + y^2) &= 0, \\
\ddot{y} + \dot{x} + y(x^2 + y^2) &= 0.
\end{aligned}$$

4. Calculate the symmetries for the following system of ODEs (a and b are nonzero constants):

$$\left(a^2 + b^2\right)\ddot{x} = (ax + by)\,\dot{x} + \dot{y},$$
$$\left(a^2 + b^2\right)\ddot{y} = \dot{x} + (ax + by)\,\dot{y}.$$

5. Calculate the symmetries for the following system of ODEs arising from a reduction of the Burgers' system (see Arrigo *et al.* [7]):

$$y\ddot{x} + xy\dot{x} - x^2\dot{y} = 0,$$
$$x\ddot{y} + xy\dot{y} - y^2\dot{x} = 0.$$

CHAPTER 3

Partial Differential Equations

In this chapter, we switch our focus to the symmetry analysis of partial differential equation (PDEs). We start with first-order equations.

3.1 FIRST-ORDER EQUATIONS

Consider equations of the form

$$F(t, x, u, u_t, u_x) = 0. \tag{3.1}$$

Extending transformations to PDEs, we seek invariance of equation (3.1) under the Lie group of infinitesimal transformations

$$\begin{aligned} \bar{t} &= t + \varepsilon T(t, x, u) + O(\varepsilon^2), \\ \bar{x} &= x + \varepsilon X(t, x, u) + O(\varepsilon^2), \\ \bar{u} &= u + \varepsilon U(t, x, u) + O(\varepsilon^2). \end{aligned} \tag{3.2}$$

As the PDE (3.1) contains first-order derivatives u_t and u_x, it is necessary to obtain extended transformations for these. In analogy with

Symmetry Analysis of Differential Equations: An Introduction,
First Edition. Daniel J. Arrigo.

ordinary differential equations we define $\overline{u}_{\overline{t}}$ and $\overline{u}_{\overline{x}}$ as

$$\overline{u}_{\overline{t}} = u_t + \varepsilon U_{[t]} + O(\varepsilon^2), \tag{3.3a}$$

$$\overline{u}_{\overline{x}} = u_x + \varepsilon U_{[x]} + O(\varepsilon^2). \tag{3.3b}$$

Similarly, we introduce the total differential operators D_t and D_x where

$$D_t = \frac{\partial}{\partial t} + u_t \frac{\partial}{\partial u} + u_{tt} \frac{\partial}{\partial u_t} + u_{tx} \frac{\partial}{\partial u_x} + \cdots, \tag{3.4a}$$

$$D_x = \frac{\partial}{\partial x} + u_x \frac{\partial}{\partial u} + u_{tx} \frac{\partial}{\partial u_t} + u_{xx} \frac{\partial}{\partial u_x} + \cdots. \tag{3.4b}$$

For the derivation of (3.3), we only consider the construction of the extended transformation for u_t as the analysis for u_x follows similarly. Therefore,

$$\begin{aligned}
\overline{u}_{\overline{t}} &= \frac{\partial(\overline{u}, \overline{x})}{\partial(\overline{t}, \overline{x})} = \frac{\partial(\overline{u}, \overline{x})}{\partial(t, x)} \bigg/ \frac{\partial(\overline{t}, \overline{x})}{\partial(t, x)} \\
&= \begin{vmatrix} \overline{u}_t & \overline{u}_x \\ \overline{x}_t & \overline{x}_x \end{vmatrix} \bigg/ \begin{vmatrix} \overline{t}_t & \overline{t}_x \\ \overline{x}_t & \overline{x}_x \end{vmatrix} \\
&= \frac{\begin{vmatrix} u_t + \varepsilon D_t(U) + O(\varepsilon^2) & u_x + \varepsilon D_x(U) + O(\varepsilon^2) \\ \varepsilon D_t(X) + O(\varepsilon^2) & 1 + \varepsilon D_x(X) + O(\varepsilon^2) \end{vmatrix}}{\begin{vmatrix} 1 + \varepsilon D_t(T) + O(\varepsilon^2) & \varepsilon D_x(T) + O(\varepsilon^2) \\ \varepsilon D_t(X) + O(\varepsilon^2) & 1 + \varepsilon D_x(X) + O(\varepsilon^2) \end{vmatrix}} \\
&= \frac{u_t + \varepsilon \left[D_t(U) + u_t D_x(X) - u_x D_t(X) \right] + O(\varepsilon^2)}{1 + \varepsilon \left[D_t(T) + D_x(X) \right] + O(\varepsilon^2)} \\
&= u_t + \varepsilon \left[D_t(U) - u_t D_t(T) - u_x D_t(X) \right] + O(\varepsilon^2).
\end{aligned}$$

Thus, we define $U_{[t]}$ and $U_{[x]}$ as

$$U_{[t]} = D_t(U) - u_t D_t(T) - u_x D_t(X), \tag{3.5a}$$

$$U_{[x]} = D_x(U) - u_t D_x(T) - u_x D_x(X). \tag{3.5b}$$

For the invariance of equation (3.1), we again introduce the infinitesimal operator Γ as

$$\Gamma = T\frac{\partial}{\partial t} + X\frac{\partial}{\partial x} + U\frac{\partial}{\partial u} \tag{3.6}$$

with its first extension as

$$\Gamma^{(1)} = \Gamma + U_{[t]}\frac{\partial}{\partial u_t} + U_{[x]}\frac{\partial}{\partial u_x}.$$

For PDEs in the form (3.1), if we denote Δ as

$$\Delta = F(t, x, u, u_t, u_x) = 0,$$

then Lie's invariance condition becomes

$$\Gamma^{(1)}\Delta\Big|_{\Delta=0} = 0 \tag{3.7}$$

as we have seen for ordinary differential equations. At this point, we are ready to consider an example.

EXAMPLE 3.1

We now consider the equation

$$u_t = u_x^2. \tag{3.8}$$

We define Δ as

$$\Delta = u_t - u_x^2.$$

Lie's invariance condition (3.7) therefore becomes

$$U_{[t]} - 2u_x U_{[x]} = 0. \tag{3.9}$$

Substituting the extensions (3.5) into (3.9) gives rise to

$$\begin{aligned} &U_t + u_t U_u - u_t(T_t + u_t T_u) - u_x(X_t + u_t X_u) \\ &\quad - 2u_x \left(U_x + u_x U_u - u_t(T_x + u_x T_u) - u_x(X_x + u_x X_u)\right) = 0. \end{aligned}$$

Substituting the original equation (3.8) and in particular eliminating u_t gives

$$\begin{aligned} &U_t - \left(X_t + 2U_x\right) u_x + \left(2X_x - T_t - U_u\right) u_x^2 \\ &\quad + \left(X_u + 2T_x\right) u_x^3 + T_u u_x^4 = 0. \end{aligned} \tag{3.10}$$

Since T, X, and U are only functions of t, x, and u, then for equation (3.10) to be satisfied requires setting the coefficients of u_x to zero. This gives rise to the following determining equations

$$U_t = 0, \tag{3.11a}$$

$$T_u = 0, \tag{3.11b}$$

$$X_t + 2U_x = 0, \tag{3.11c}$$

$$X_u + 2T_x = 0, \tag{3.11d}$$

$$2X_x - T_t - U_u = 0. \tag{3.11e}$$

From (3.11b), we obtain

$$T = A(t, x), \tag{3.12}$$

where A is an arbitrary function. Substituting (3.12) into (3.11d) and integrating gives

$$X = -2A_x u + B. \tag{3.13}$$

where $B = B(t, x)$ is another arbitrary function. Substitution into (3.11e) gives

$$U_u + A_t + 4A_{xx}u - 2B_x = 0,$$

which integrates to give

$$U = -2A_{xx}u^2 + \left(2B_x - A_t\right) u + C, \tag{3.14}$$

where $C = C(t, x)$ is again an arbitrary function. Substituting (3.12), (3.13), and (3.14) into the remaining equations of (3.11) and isolating the coefficients of u gives rise to

$$\begin{array}{lll} A_{xxx} = 0, & A_{tx} - B_{xx} = 0, & B_t + 2C_x = 0, \\ A_{txx} = 0, & A_{tt} - 2B_{tx} = 0, & C_t = 0. \end{array} \tag{3.15}$$

Solving the system of equations (3.15) gives rise to

$$\begin{aligned} A &= c_1 x^2 + (2c_2 t + c_3)x + 4c_4 t^2 + c_5 t + c_6 \\ B &= c_2 x^2 + (4c_4 t + c_7)x + 2c_8 t + c_9 \\ C &= -c_4 x^2 - c_8 x + c_{10}, \end{aligned}$$

which in turn give rise to the infinitesimals T, X, and U. These are given by

$$\begin{aligned} T &= c_1 x^2 + (2c_2 t + c_3)x + 4c_4 t^2 + c_5 t + c_6, \\ X &= -2(2c_1 x + 2c_2 t + c_3)u + c_2 x^2 + (4c_4 t + c_7)x + 2c_8 t + c_9, \\ U &= -4c_1 u^2 + (2c_2 x - c_5 + 2c_7)u - c_4 x^2 - c_8 x + c_{10}, \end{aligned} \tag{3.16}$$

where $c_1 - c_{10}$ are arbitrary constants. The natural question is now: what do we do with these infinitesimals? With ordinary differential equations, we introduced new variables, r and s, such that the original equation reduced to a new ordinary differential equation that was separable equation and independent of the variable s. Do these ideas also apply to partial differential equations? ■

3.1.1 What Do We Do with the Symmetries of PDEs?

With ordinary differential equations invariant under the transformation

$$\bar{x} = x + \varepsilon X(x, y) + O(\varepsilon^2),$$
$$\bar{y} = y + \varepsilon Y(x, y) + O(\varepsilon^2), \tag{3.17}$$

we found a change of variables by solving the system of equations

$$Xr_x + Yr_y = 0, \quad Xs_x + Ys_y = 1. \tag{3.18}$$

With the introduction of these new variables r and s, the original equation is transformed to one that is independent of s. To illustrate this fact further, consider the infinitesimal operator Γ associated with (3.17)

$$\Gamma = X\frac{\partial}{\partial x} + Y\frac{\partial}{\partial y}. \tag{3.19}$$

Under a change of variables $(x, y) \to (r, s)$, the operator (3.19) becomes

$$\begin{aligned}
\Gamma &= X\frac{\partial}{\partial x} + Y\frac{\partial}{\partial y} \\
&= X\left(r_x\frac{\partial}{\partial r} + s_x\frac{\partial}{\partial s}\right) + Y\left(r_y\frac{\partial}{\partial r} + s_y\frac{\partial}{\partial s}\right) \\
&= \left(Xr_x + Yr_y\right)\frac{\partial}{\partial r} + \left(Xs_x + Ys_y\right)\frac{\partial}{\partial s}
\end{aligned}$$

and imposing the condition (3.18) gives

$$\Gamma = \frac{\partial}{\partial s},$$

which represents that the new ODE admits translation in the variable s, that is, the new ODE is independent of s.

We now attempt to extend these ideas to PDEs. Consider the infinitesimal operator associated with the infinitesimal transformations (3.2)

$$\Gamma = T\frac{\partial}{\partial t} + X\frac{\partial}{\partial x} + U\frac{\partial}{\partial u}. \tag{3.20}$$

If we introduce new independent variables r and s and a new dependent variable v such that

$$r = r(t, x, u), \qquad s = s(t, x, u), \qquad v = v(t, x, u),$$

then under a change of variables $(t, x, u) \to (r, s, v)$, the operator (3.20) becomes

$$\begin{aligned}
\Gamma &= T\frac{\partial}{\partial t} + X\frac{\partial}{\partial x} + U\frac{\partial}{\partial u},\\
&= T\left(r_t\frac{\partial}{\partial r} + s_t\frac{\partial}{\partial s} + v_t\frac{\partial}{\partial v}\right) + X\left(r_x\frac{\partial}{\partial r} + s_x\frac{\partial}{\partial s} + v_x\frac{\partial}{\partial v}\right)\\
&\quad + U\left(r_u\frac{\partial}{\partial r} + s_u\frac{\partial}{\partial s} + v_u\frac{\partial}{\partial v}\right)\\
&= \left(Tr_t + Xr_x + Ur_u\right)\frac{\partial}{\partial r} + \left(Ts_t + Xs_x + Us_u\right)\frac{\partial}{\partial s}\\
&\quad + \left(Tv_t + Xv_x + Uv_u\right)\frac{\partial}{\partial v},
\end{aligned} \tag{3.21}$$

which, upon choosing

$$\begin{aligned}
Tr_t + Xr_x + Ur_u &= 0,\\
Ts_t + Xs_x + Us_u &= 1,\\
Tv_t + Xv_x + Uv_u &= 0,
\end{aligned} \tag{3.22}$$

in (3.21) gives

$$\Gamma = \frac{\partial}{\partial s}.$$

This says that after a change of variables, the transformed equation will be independent of s. So what does this mean? If the original equation has the form

$$F(t, x, u, u_t, u_x) = 0,$$

then the transformed equation will be of the form

$$G(r, v, v_r, v_s) = 0. \tag{3.23}$$

The transformed equation is still a PDE; however, if we assume that we are only interested in solutions of the form $v = v(r)$, then (3.23) becomes

$$H(r, v, v_r) = 0,$$

an ODE!

EXAMPLE 3.2

We again revisit equation (3.8) where we constructed the symmetries of

$$u_t = u_x^2.$$

These were obtained in (3.16). As a particular example, consider the case where $c_5 = 1$ and $c_7 = 1$ with the remaining constants zero. This gives

$$T = t, \qquad X = x, \qquad U = u.$$

The change of variable equations (3.22) for the new variables r, s, and v is

$$\begin{aligned} tr_t + xr_x + ur_u &= 0, \\ ts_t + xs_x + us_u &= 1, \\ tv_t + xv_x + uv_u &= 0. \end{aligned}$$

The solution is given by

$$r = R\left(\frac{x}{t}, \frac{u}{t}\right), \quad s = \ln t + S\left(\frac{x}{t}, \frac{u}{t}\right), \quad v = V\left(\frac{x}{t}, \frac{u}{t}\right),$$

where R, S, and T are arbitrary functions of their arguments. If we choose these to give

$$r = \frac{x}{t}, \quad s = \ln t, \quad v = \frac{u}{t}, \tag{3.24}$$

then transforming the original equation (3.8) gives rise to

$$v + v_s - rv_r = v_r^2.$$

Setting the term $v_s = 0$ and simplifying gives[1]

$$v_r^2 + rv_r - v = 0,$$

[1] This particular ODE is known as Clairaut equation whose general form is

$$y = xy' + f(y').$$

whose general solution is

$$v = -\frac{1}{4}r^2, \quad v = cr + c^2.$$

where c is an arbitrary constant. Passing through the transformation (3.24) gives rise to the following exact solutions to (3.8)

$$u = -\frac{1}{4}\frac{x^2}{t}, \quad u = cx + c^2 t.$$

By choosing different constants in the infinitesimals (3.16), we undoubtedly will obtain new transformed PDEs, from which we would obtain ODEs to solve which could lead to new exact solutions to the original equation. A natural question is: can we bypass the stage of introducing the new variables r, s, and v and go directly to the ODE? ■

3.1.2 Direct Reductions

In the previous section, we considered the invariance of

$$\Delta = u_t - u_x^2,$$

and we obtained infinitesimals T, X, and U leaving the original equation invariant. We introduced new variables r, s, and v transforming the equation to a new equation independent of s. Setting $v_s = 0$ gave rise to an ODE which we solved giving rise to an exact solution to the original PDE. In this section, we show that we can go directly to the ODE without the intermediate step of introducing new variables r, s, and v. Recall that the new variables were found by solving

$$Tr_t + Xr_x + Ur_u = 0,$$
$$Ts_t + Xs_x + Us_u = 1,$$
$$Tv_t + Xv_x + Uv_u = 0.$$

Solving by the method of characteristics give

$$\frac{dt}{T} = \frac{dx}{X} = \frac{du}{U}; \quad dr = 0,$$
$$\frac{dt}{T} = \frac{dx}{X} = \frac{du}{U} = \frac{ds}{1},$$
$$\frac{dt}{T} = \frac{dx}{X} = \frac{du}{U}; \quad dv = 0.$$

From the characteristic equations, it suffices to only choose

$$\frac{\mathrm{d}t}{T} = \frac{\mathrm{d}x}{X} = \frac{\mathrm{d}u}{U}. \tag{3.25}$$

Once this is solved, this will give the solution form or "ansatz" and upon substitution into the original equation, we will obtain an ODE that will then need to be solved. The following example illustrates.

EXAMPLE 3.3

In Example 3.2, we considered the PDE

$$u_t = u_x^2,$$

in which we calculated the infinitesimals T, X, and U given in (3.16). One particular example was

$$T = t, \qquad X = x, \qquad U = u.$$

To go directly to the solution ansatz, we solve (3.25), namely

$$\frac{\mathrm{d}t}{t} = \frac{\mathrm{d}x}{x} = \frac{\mathrm{d}u}{u}.$$

The solution is given by

$$u = tF\left(\frac{x}{t}\right). \tag{3.26}$$

Substituting (3.26) into (3.8) gives

$$F - rF' = F'^2,$$

where $F = F(r)$ and $r = \frac{x}{t}$. This is exactly the ODE we obtained in the previous section. Let us consider another example. ■

EXAMPLE 3.4

Choosing

$$T = x^2, \qquad X = -4xu, \qquad U = -4u^2,$$

the characteristic equation (3.25) becomes

$$\frac{\mathrm{d}t}{x^2} = \frac{\mathrm{d}x}{-4xu} = \frac{\mathrm{d}u}{-4u^2}.$$

The solution is given by

$$x + \frac{4tu}{x} = F\left(\frac{u}{x}\right) \tag{3.27}$$

noting that in this case, the solution is implicit. Substituting (3.27) into (3.8) (with $r = u/x$) gives

$$(rF'(r) - F(r))^2 = 0, \quad \text{or} \quad rF' - F = 0,$$

which remarkably is linear! Solving gives

$$F(r) = kr,$$

where k is an arbitrary constant. This leads, via equation (3.27) to

$$x + \frac{4tu}{x} = k\frac{u}{x}$$

which we solve for u leads to

$$u = \frac{x^2}{k - 4t}.$$

■

EXAMPLE 3.5

Choosing the constants in the infinitesimals (3.16)

$$c_6 = 1, \quad c_8 = -3,$$

and the rest zero gives rise to

$$T = 1, \qquad X = -6t, \qquad U = 3x.$$

Solving the characteristic equation (3.25) with these infinitesimals

$$\frac{dt}{1} = \frac{dx}{-6t} = \frac{du}{3x}$$

gives the solution

$$u = 3xt + 6t^3 + F\left(x + 3t^2\right).$$

Substituting into (3.8) gives

$$F'^2 = 3r,$$

where $r = x + 3t^2$. Solving gives

$$F(r) = \frac{2\sqrt{3}}{3} r^{3/2} + c,$$

where c is an arbitrary constant. This leads to the exact solution

$$u = 3xt + 6t^3 + \frac{2\sqrt{3}}{3}(x + 3t^2)^{3/2} + c.$$

■

3.1.3 The Invariant Surface Condition

We have seen that we can go directly to the solution ansatz by solving the characteristics equation

$$\frac{\mathrm{d}t}{T} = \frac{\mathrm{d}x}{X} = \frac{\mathrm{d}u}{U}.$$

With this, we can associate the first-order PDE

$$Tu_t + Xu_x = U. \tag{3.28}$$

This is usually referred to as the "invariant surface condition." The reason for this name is because the solution surface remains invariant under the infinitesimal transformation. To demonstrate, consider the solution

$$u = u(t, x).$$

If we consider

$$\bar{u} = u(\bar{t}, \bar{x}) \tag{3.29}$$

then under the infinitesimal transformations

$$\bar{t} = t + \varepsilon T(t, x, u) + O(\varepsilon^2),$$
$$\bar{x} = x + \varepsilon X(t, x, u) + O(\varepsilon^2),$$
$$\bar{u} = u + \varepsilon U(t, x, u) + O(\varepsilon^2).$$

Equation (3.29) becomes

$$u + \varepsilon U(t, x, u) + O(\varepsilon^2) = u(t + \varepsilon T(t, x, u) + O(\varepsilon^2), x + \varepsilon X(t, x, u) + O(\varepsilon^2)),$$

and expanding gives

$$u + \varepsilon U(t,x,u) + O(\varepsilon^2) = u(t,x) + \varepsilon \left(T(t,x,u)u_t + X(t,x,u)u_x \right) + O(\varepsilon^2).$$

If $u = u(t,x)$, then to order ε, we obtain

$$Tu_t + Xu_x = U,$$

the invariant surface condition (equation (3.28)).

EXERCISE

1. Calculate the symmetries for the the first-order PDEs ($c \in \mathbb{R}$)

$$\begin{array}{lrcl} \text{(i)} & u_t + cu_x & = & 0 \\ \text{(ii)} & u_t + uu_x & = & 0 \\ \text{(iii)} & u_t + u_x^2 + u & = & 0 \\ \text{(iv)} & xu_xu_y + yu_y^2 & = & 1. \end{array}$$

Use a particular symmetry to reduce the PDE to an ODE and solve if possible.

3.2 SECOND-ORDER PDEs

3.2.1 Heat Equation

In this section, we consider constructing the symmetries of the heat equation

$$u_t = u_{xx}. \tag{3.30}$$

In analogy with second-order ODEs, we define the second extension $\Gamma^{(2)}$ of the operator Γ. If Γ is given by

$$\Gamma = T\frac{\partial}{\partial t} + X\frac{\partial}{\partial x} + U\frac{\partial}{\partial u},$$

then the first and second extensions are given by

$$\Gamma^{(1)} = \Gamma + U_{[t]}\frac{\partial}{\partial u_t} + U_{[x]}\frac{\partial}{\partial u_x}, \tag{3.31a}$$

$$\Gamma^{(2)} = \Gamma^{(1)} + U_{[tt]}\frac{\partial}{\partial u_{tt}} + U_{[tx]}\frac{\partial}{\partial u_{tx}} + U_{[xx]}\frac{\partial}{\partial u_{xx}}, \tag{3.31b}$$

respectively. The extended transformations are given as

$$U_{[t]} = D_t(U) - u_t D_t(T) - u_x D_t(X), \tag{3.32a}$$

$$U_{[x]} = D_x(U) - u_t D_x(T) - u_x D_x(X), \tag{3.32b}$$

$$U_{[tt]} = D_t(U_{[t]}) - u_{tt} D_t(T) - u_{tx} D_t(X), \tag{3.32c}$$

$$\begin{aligned} U_{[tx]} &= D_t(U_{[x]}) - u_{tx} D_t(T) - u_{xx} D_t(X), \\ &= D_x(U_{[t]}) - u_{tt} D_x(T) - u_{tx} D_x(X), \end{aligned} \tag{3.32d}$$

$$U_{[xx]} = D_x(U_{[x]}) - u_{tx} D_x(T) - u_{xx} D_x(X), \tag{3.32e}$$

where D_t and D_x are given in (3.4). We expand only $U_{[xx]}$ to demonstrate

$$\begin{aligned} U_{[xx]} = {} & U_{xx} - T_{xx}u_t + \left(2U_{xu} - X_{xx}\right)u_x - 2T_{xu}u_t u_x \\ & + \left(U_{uu} - 2X_{xu}\right)u_x^2 - T_{uu}u_t u_x^2 - X_{uu}u_x^3 \\ & -2T_x u_{tx} + \left(U_u - 2X_x\right)u_{xx} - 2T_u u_x u_{tx} \\ & -T_u u_t u_{xx} - 3X_u u_x u_{xx}. \end{aligned}$$

If we denote the heat equation (3.30) as Δ, then

$$\Delta = u_t - u_{xx}.$$

As with second-order ODEs, Lie's invariance condition is

$$\Gamma^{(2)}\Delta\Big|_{\Delta=0} = 0,$$

however, we use the extended operator as given in (3.31). This, in turn, gives

$$U_{[t]} - U_{[xx]} = 0. \tag{3.33}$$

Substituting the extended transformations (3.32a) and (3.32e) into (3.33) gives rise to

$$\begin{aligned} & U_t + \left(U_u - T_t\right)u_t - X_t u_x - T_u u_t^2 - X_u u_t u_x \\ & - U_{xx} + T_{xx}u_t - \left(2U_{xu} - X_{xx}\right)u_x + 2T_{xu}u_t u_x \\ & - \left(U_{uu} - 2X_{xu}\right)u_x^2 + T_{uu}u_t u_x^2 + X_{uu}u_x^3 \end{aligned}$$

$$+2T_x u_{tx} - \left(U_u - 2X_x\right) u_{xx} + 2T_u u_x u_{tx}$$
$$+T_u u_t u_{xx} + 3X_u u_x u_{xx} = 0.$$

Using the heat equation to eliminate u_{xx} and isolating coefficients involving u_t, u_x, and u_{tx} and products give rise to the following set of determining equations:

$$U_t - U_{xx} = 0, \tag{3.34a}$$
$$-X_t - 2U_{xu} + X_{xx} = 0, \tag{3.34b}$$
$$-U_{uu} + 2X_{xu} = 0, \tag{3.34c}$$
$$X_{uu} = 0, \tag{3.34d}$$
$$-T_t + T_{xx} + 2X_x = 0, \tag{3.34e}$$
$$2X_u + 2T_{xu} = 0, \tag{3.34f}$$
$$T_{uu} = 0, \tag{3.34g}$$
$$2T_x = 0, \tag{3.34h}$$
$$2T_u = 0. \tag{3.34i}$$

We find immediately from (3.34) that

$$T_x = 0, \quad T_u = 0, \quad X_u = 0, \quad U_{uu} = 0, \tag{3.35}$$

leaving

$$2X_x - T_t = 0, \tag{3.36a}$$
$$-2U_{xu} + X_{xx} - X_t = 0, \tag{3.36b}$$
$$U_t - U_{xx} = 0. \tag{3.36c}$$

From (3.35) and (3.36a) that

$$X = \frac{1}{2}T'(t)x + A(t), \tag{3.37a}$$
$$U = B(t,x)u + C(t,x), \tag{3.37b}$$

where A, B, and C are arbitrary functions. Substituting (3.37) into (3.36b) gives

$$2B_x + \frac{1}{2}T''x + A' = 0,$$

which integrates to give

$$B = -\frac{1}{8}T''(t)x^2 - \frac{1}{2}A'(t)x + D(t), \tag{3.38}$$

where D is a function of integration. Finally, substitution of (3.37b) and (3.38) into (3.36c) gives

$$\left(-\frac{1}{8}T'''(t)x^2 - \frac{1}{2}A''(t)x + D'(t) + \frac{1}{4}T''(t)\right)u + C_t - C_{xx} = 0.$$

Isolating coefficients of x and u gives

$$T''' = 0, \quad A'' = 0, \quad D' + \frac{1}{4}T'' = 0, \quad C_t - C_{xx} = 0.$$

Solving gives the following infinitesimals:

$$\begin{aligned} T &= c_0 + 2c_1 t + 4c_2 t^2, \\ X &= c_3 + 2c_4 t + c_1 x + 4c_2 tx, \\ U &= \left(c_5 - 2c_2 t - c_4 x - c_2 x^2\right) u + C(t,x), \end{aligned} \tag{3.39}$$

where $c_1 - c_5$ are constant and C satisfies the heat equation.

Reductions of the Heat Equation

In this section, we construct exact solutions of the heat equation via symmetry analysis. To do this, we consider the invariant surface condition (3.28) introduced in an earlier section

$$Tu_t + Xu_x = U. \tag{3.40}$$

As there are a total of five arbitrary constants in addition to an arbitrary function in (3.39), we only consider a few examples.

EXAMPLE 3.6

If we choose $c_1 = 1$ and $c_i = 0$ otherwise and $C(t,x) = 0$ gives

$$T = 2t, \qquad X = x, \qquad U = 0.$$

This, in turn, leads us to consider the invariant surface (3.40) condition

$$2tu_t + xu_x = 0,$$

whose solution is given by

$$u = F\left(\frac{x}{\sqrt{t}}\right).$$

Substituting into the heat equation gives

$$-\frac{r}{2}F'(r) = F''(r), \qquad r = \frac{x}{\sqrt{t}}.$$

Integrating once gives

$$F'(r) = k_1 e^{-r^2/4}$$

and further integration, we obtain

$$F(r) = k_1 \text{erf}(r) + k_2,$$

where k_1 and k_2 are arbitrary constants and erf(x) is the error function as defined by $\text{erf}(x) = \int_{-\infty}^{x} e^{-\xi^2/4} d\xi$. In terms of the original variables, this gives rise to the exact solution

$$u = k_1 \text{erf}\left(\frac{x}{\sqrt{t}}\right) + k_2.$$

■

EXAMPLE 3.7

If we choose $c_2 = 1$ and $c_i = 0$ otherwise and $C(t, x) = 0$ gives the invariant surface condition (3.40)

$$4t^2 u_t + 4txu_x = -(2t + x^2)u,$$

whose solution is given by

$$u = \frac{e^{-x^2/4t}}{\sqrt{t}} F\left(\frac{x}{t}\right).$$

Substituting into the heat equation gives

$$F''(r) = 0, \qquad r = \frac{x}{\sqrt{t}}.$$

Integrating gives

$$F(r) = k_1 r + k_2,$$

where k_1 and k_2 are arbitrary constants. This gives rise to the exact solution

$$u = k_1 \frac{x}{t\sqrt{t}} \, \mathrm{e}^{-x^2/4t} + k_2 \frac{1}{\sqrt{t}} \, \mathrm{e}^{-x^2/4t}.$$

■

EXAMPLE 3.8

If we choose $c_4 = 1$ and $c_i = 0$ otherwise and $C(t, x) = 0$ gives the invariant surface condition (3.40)

$$2tu_x = -xu,$$

whose solution is given by

$$u = e^{-x^2/4t} F(t).$$

Substituting into the heat equation gives

$$F' + \frac{F}{2t} = 0.$$

Integrating gives

$$F(t) = \frac{k_1}{\sqrt{t}}$$

giving rise to the exact solution

$$u = k_1 \, \frac{\mathrm{e}^{-x^2/4t}}{\sqrt{t}}.$$

It is important to point out that this solution was also obtained in the previous case demonstrating that different symmetries may lead to the same solution. ■

Generating New Solutions

In the preceding three examples, we set $C(t, x) = 0$ and used special cases of the infinitesimals to obtain exact solutions through symmetry reduction. If we consider the case where $C(t, x) \neq 0$, then the invariant surface condition is

$$Tu_t + Xu_x = Bu + C.$$

If we isolate C, then we have

$$C = Tu_t + Xu_x - Bu.$$

It is an easy matter to verify that

$$C_t = C_{xx},$$

if u satisfies the heat equation and T, X, and B are

$$\begin{aligned} T &= c_0 + 2c_1 t + 4c_2 t^2, \\ X &= c_3 + 2c_4 t + c_1 x + 4c_2 tx, \\ B &= c_5 - 2c_2 t - c_4 x - c_2 x^2. \end{aligned}$$

This means that it is possible to generate a hierarchy of exact solutions given a "seed" solution. For example, if we consider

$$C = 4t^2 u_t + 4txu_x + (2t + x^2)u, \tag{3.41}$$

the infinitesimals used in Example 3.7 and use the exact solution

$$u = \operatorname{erf}\left(\frac{x}{2\sqrt{t}}\right)$$

as obtained in Example 3.6, then from (3.41) we obtain the new exact solution

$$u = \left(x^2 + 2t\right) \operatorname{erf}\left(\frac{x}{2\sqrt{t}}\right) + \frac{2x\sqrt{t}}{\sqrt{\pi}} \mathrm{e}^{-x^2/4t}.$$

We could easily choose different seed solutions or repeat the process to obtain an entire hierarchy of exact solutions.

3.2.2 Laplace's Equation

In this section, we consider the invariance of Laplace's equation

$$u_{xx} + u_{yy} = 0. \tag{3.42}$$

For this equation, Lie's invariance condition becomes

$$U_{[xx]} + U_{[yy]} = 0, \tag{3.43}$$

where the extended transformations are given as

$$\begin{aligned}
U_{[x]} &= D_x(U) - u_x D_x(X) - u_y D_x(Y),\\
U_{[y]} &= D_y(U) - u_x D_y(X) - u_y D_y(Y),\\
U_{[xx]} &= D_x(U_{[x]}) - u_{xx} D_x(X) - u_{xy} D_x(Y),\\
U_{[yy]} &= D_y(U_{[y]}) - u_{xy} D_y(X) - u_{yy} D_y(Y).
\end{aligned}$$

Expanding Lie's invariance condition (3.43) gives

$$\begin{aligned}
&U_{xx} + U_{yy} + (2U_{xu} - X_{xx} - X_{yy})u_x + (2U_{yu} - Y_{xx} - Y_{yy})u_y\\
&\quad +(U_{uu} - 2X_{xu})u_x^2 - 2(X_{yu} + Y_{xu})u_x u_y + (U_{uu} - 2Y_{yu})u_y^2\\
&\quad -X_{uu}u_x^3 - Y_{uu}u_x^2 u_y - X_{uu}u_x u_y^2 - Y_{uu}u_y^3 + (U_u - 2X_x)u_{xx}\\
&\quad -2(X_y + Y_x)u_{xy} + (U_u - 2Y_y)u_{yy} - 3X_u u_x u_{xx} - Y_u u_y u_{xx}\\
&\quad -2Y_u u_x u_{xy} - 2X_u u_y u_{xy} - X_u u_x u_{yy} - 3Y_u u_y u_{yy} = 0.
\end{aligned}$$

Substituting in the original equation $u_{xx} = -u_{yy}$ (or $u_{yy} = -u_{xx}$) and setting the coefficients to zero of the terms that involve u_x, u_y, u_{xy} and u_{yy} (or u_{xx}) and various products gives

$$U_{xx} + U_{yy} = 0, \tag{3.44a}$$

$$2U_{xu} - X_{xx} - X_{yy} = 0, \tag{3.44b}$$

$$2U_{yu} - Y_{xx} - Y_{yy} = 0, \tag{3.44c}$$

$$U_{uu} - 2X_{xu} = 0, \tag{3.44d}$$

$$X_{yu} + Y_{xu} = 0, \tag{3.44e}$$

$$U_{uu} - 2Y_{yu} = 0, \tag{3.44f}$$

$$X_{uu} = Y_{uu} = 0, \tag{3.44g}$$

$$Y_y - X_x = 0, \tag{3.44h}$$

$$X_y + Y_x = 0, \tag{3.44i}$$

$$X_u = Y_u = 0. \tag{3.44j}$$

As $X_u = 0$ and $Y_u = 0$ then $U_{uu} = 0$, giving

$$X = A(x, y), \qquad Y = B(x, y), \qquad U = P(x, y)u + Q(x, y),$$

where A, B, P and Q are arbitrary functions. From (3.44h) and (3.44i), A and B satisfy

$$A_x - B_y = 0, \qquad A_y + B_x = 0. \tag{3.45}$$

Furthermore, from (3.44b) and (3.44c), we see that

$$P_x = 0, \qquad P_y = 0,$$

from which it follows that

$$P(x, y) = p,$$

where p is an arbitrary constant. Finally, from (3.44a) we have that Q satisfies

$$Q_{xx} + Q_{yy} = 0. \tag{3.46}$$

This gives the infinitesimals as

$$X = A(x, y), \qquad Y = B(x, y), \qquad U = p\,u + Q(x, y),$$

where A and B satisfy (3.45) and Q Laplace's equation (3.46). We now construct some exact solutions using these infinitesimals.

EXAMPLE 3.9

A particular solution to the equations (3.45) and (3.46) is

$$A = x, \qquad B = y, \qquad Q = 0.$$

The associated invariant surface condition is

$$xu_x + yu_y = pu,$$

TABLE 3.1
Real solutions of equation (3.48) for some integer values of p

p	F_1	F_2	p	F_1	F_2
1	1	r	−1	$\frac{1}{r^2+1}$	$\frac{r}{r^2+1}$
2	r	r^2-1	−2	$\frac{r}{(r^2+1)^2}$	$\frac{r^2-1}{(r^2+1)^2}$
3	$3r^2-1$	r^3-3r	−3	$\frac{3r^2-1}{(r^2+1)^3}$	$\frac{r^3-3r}{(r^2+1)^3}$
4	r^3+r	r^4-6r^2+1	−4	$\frac{r^3+r}{(r^2+1)^4}$	$\frac{r^4-6r^2+1}{(r^2+1)^4}$

which is easily solved giving

$$u = x^p F\left(\frac{y}{x}\right), \tag{3.47}$$

where F is arbitrary. Substitution of (3.47) into Laplace's equation (3.42) gives

$$(r^2+1)F'' - 2(p-1)rF' + (p^2-p)F = 0, \qquad r = \frac{y}{x}. \tag{3.48}$$

This can be integrated exactly giving

$$F = \begin{cases} c_1 \tan^{-1} r + c_2, & \text{if } p = 0 \\ c_1(r-i)^p + c_2(r+i)^p & \text{if } p \neq 0 \end{cases},$$

where $i = \sqrt{-1}$. If $p \neq 0$, Table 3.1 presents two independent real solutions F_1 and F_2 for a variety of integer powers p

If $p = 0$, then we obtain the exact solution

$$u = c_1 \tan^{-1} \frac{y}{x} + c_2.$$

If $p \neq 0$, exact solutions are then given by

$$u = x^p \left[c_1 F_1\left(\frac{y}{x}\right) + c_2 F_2\left(\frac{y}{x}\right)\right].$$

where F_1 and F_2 are real solutions of (3.48) and c_1 and c_2 new constants. ■

EXAMPLE 3.10

Another particular solution to the equations (3.45) and (3.46) is

$$A = y, \qquad B = -x, \qquad P = Q = 0.$$

The associated invariant surface condition is

$$yu_x - xu_y = 0,$$

which is easily solved giving

$$u = F\left(x^2 + y^2\right),$$

where F is arbitrary. Substitution into Laplaces equation gives

$$rF'' + F' = 0, \qquad r = x^2 + y^2,$$

which is easily solved giving

$$F = c_1 \ln r + c_2,$$

leading to the exact solution

$$u = c_1 \ln\left(x^2 + y^2\right) + c_2.$$

■

3.2.3 Burgers' Equation and a Relative

In this section, we consider constructing the symmetries of Burgers' equation

$$u_t + uu_x = u_{xx} \tag{3.49}$$

and its relative, the potential Burgers' equation

$$v_t + \frac{1}{2}v_x^2 = v_{xx}. \tag{3.50}$$

These two equations are linked by $u = v_x$. We consider the symmetry analysis of each separately.

Burgers' Equation

Lie's invariance condition (3.49) leads to

$$U_{[t]} + uU_{[x]} + Uu_x = U_{[xx]}.$$

Substituting the appropriate extended transformations eventually leads to the following set of determining equations:

$$\begin{gathered} T_x = 0, \quad T_u = 0, \quad X_u = 0, \quad U_{uu} = 0, \\ 2X_x - T_t = 0, \quad U_t + uU_x - U_{xx} = 0, \\ X_t - X_{xx} - uX_x - U + 2U_{xu} = 0. \end{gathered} \tag{3.51}$$

Solving the first 5 equations of (3.51) gives

$$X = \frac{1}{2}T'(t)x + A(t), \tag{3.52a}$$

$$U = B(t,x)u + C(t,x), \tag{3.52b}$$

where $A(t)$, $B(t,x)$ and $C(t,x)$ are arbitrary functions. Substituting (3.52) into the remaining two equations of (3.51), isolating coefficients with respect to u and solving gives the infinitesimals:

$$\begin{aligned} T &= c_0 + 2c_1 t + c_2 t^2, \\ X &= c_3 + c_4 t + c_1 x + c_2 t x, \\ U &= -\left(c_2 t + c_1\right) u + c_2 x + c_4, \end{aligned} \tag{3.53}$$

where c_1, c_2, c_3 and c_4 are arbitrary constants.

Potential Burgers' Equation

Lie's invariance condition of (3.50) leads to

$$V_{[t]} + v_x V_{[x]} = V_{[xx]},$$

which substituting the appropriate extended transformations eventually leads to the following set of determining equations:

$$\begin{gathered} T_x = 0, \quad T_v = 0, \quad X_v = 0, \\ 2X_x - T_t = 0, \quad V_t - V_{xx} = 0, \quad V_{vv} - \frac{1}{2}V_v = 0, \\ X_t - X_{xx} - V_x + 2V_{xv} = 0. \end{gathered}$$

Again, these are readily solved giving the infinitesimals:

$$
\begin{aligned}
T &= c_0 + 2c_1 t + c_2 t^2, \\
X &= c_3 + c_4 t + c_1 x + c_2 tx, \\
V &= P(t, x)e^{v/2} + \frac{1}{2}c_2 x^2 + c_4 x + c_2 t + c_5.
\end{aligned}
\tag{3.54}
$$

where c_1, c_2, c_3, c_4 and c_5 are arbitrary constants and $P(t, x)$ satisfies the heat equation $P_t = P_{xx}$.

Reductions

In this section, we consider examples of reductions of both Burgers' equation and the potential Burgers' equation.

EXAMPLE 3.11 $c_1 = 1, \quad c_i = 0, \quad i \neq 1$

In the case of Burgers' equation with (3.53), the invariant surface condition is

$$2tu_t + xu_x = -u,$$

which is easily solved giving

$$u = \frac{1}{\sqrt{t}} F\left(\frac{x}{\sqrt{t}}\right).$$

Substitution into (3.49) gives the reduction

$$-\frac{1}{2}F(r) - \frac{1}{2}rF'(r) + F(r)F'(r) - F''(r) = 0,$$

which we find integrates once to give the Ricatti equation

$$-\frac{1}{2}rF + F^2 - F' = c_1,$$

where c_1 is a constant of integration. In the case of the Potential Burgers' equation with (3.54), the invariant surface condition is

$$2tv_t + xv_x = 0,$$

which is easily solved giving

$$v = F\left(\frac{x}{\sqrt{t}}\right).$$

Substitution into (3.50) gives the reduction

$$-\frac{1}{2}rF'(r) + \frac{1}{2}F'^2(r) - F''(r) = 0,$$

whose solution is given by

$$F(r) = -2\ln\left(c_1 \operatorname{erf}(r/2) + c_2\right)$$

This, in turn, leads to the exact solution

$$v = -2\ln\left(c_1 \operatorname{erf}(x/2\sqrt{t}) + c_2\right).$$

■

EXAMPLE 3.12 $c_2 = 1, \quad c_i = 0, \quad i \neq 2$

In the case of Burgers' equation with (3.53), the invariant surface condition is

$$t^2u_t + xtu_x = -tu + x,$$

which is easily solved giving

$$u = \frac{x}{t} + \frac{1}{t}F\left(\frac{x}{t}\right).$$

Substitution into (3.49) gives the reduction

$$F(r)F'(r) - F''(r) = 0,$$

which integrates once to give

$$F'(r) = \frac{1}{2}F^2 + k,$$

which gives three solutions depending whether $k > 0$, $k = 0$, or $k < 0$.
In the case of the potential Burgers' equation with (3.54), the invariant surface condition is

$$t^2v_t + xtv_x = \frac{1}{2}x^2 + t,$$

which is easily solved giving

$$v = \frac{x^2}{2t} + \ln t + F\left(\frac{x}{t}\right).$$

Substitution into (3.50) gives the reduction

$$\frac{1}{2}F'^2(r) - F''(r) = 0,$$

whose solution is given by

$$F(r) = -2\ln\left(c_1 r + c_2\right)$$

This, in turn, leads to the exact solution

$$v = \frac{x^2}{2t} + \ln t - 2\ln\left|c_1\frac{x}{t} + c_2\right|.$$

■

Linearization of the Potential Burgers' Equation

In constructing the infinitesimals for both the potential Burgers' equation and the heat equation, each possessed an arbitrary function that satisfied the heat equation. This suggests that the two sets of determining equations might be linked. Below these are listed side-by-side. We have chosen to use the variable w for the heat equation as to avoid confusion with Burgers' equation.

Heat equation	Potential Burgers' equation	
$T_x = T_w = X_w = 0$	$T_x = T_v = X_v = 0$	(3.55a)
$2X_x - T_t = 0$	$2X_x - T_t = 0$	(3.55b)
$W_t - W_{xx} = 0$	$V_t - V_{xx} = 0$	(3.55c)
$2W_{xw} + X_t - X_{xx} = 0$	$2V_{xv} - V_x + X_t - X_{xx} = 0$	(3.55d)
$W_{ww} = 0$	$V_{vv} - \frac{1}{2}V_v = 0.$	(3.55e)

If there existed a transformation connecting the equations (heat and Potential Burgers'), this same transformation would connect their symmetries. Let us assume that this transformation is of the form

$$w = F(v) \tag{3.56}$$

thus,

$$W = F'(v)V. \tag{3.57}$$

It is a simple matter to deduce that

$$\left(W_t - W_{xx}\right) = F'(v)\left(V_t - V_{xx}\right)$$

thus connecting (3.55c). Furthermore, from (3.56), we find that

$$\frac{\partial}{\partial w} = \frac{1}{F'(v)}\frac{\partial}{\partial v}. \tag{3.58}$$

Thus, combining (3.57) and (3.58)

$$\begin{aligned} W_w &= \frac{1}{F'}\left(F'(v)V\right)_v \\ &= V_v + \frac{F''}{F'}V, \end{aligned} \tag{3.59}$$

which further gives

$$W_{xw} = V_{xv} + \frac{F''}{F'}V_x. \tag{3.60}$$

In order to connect (3.55d), we would require that

$$\frac{2F''}{F'} = -1. \tag{3.61}$$

Applying the chain rule (3.58) to (3.59) gives

$$\begin{aligned} W_{ww} &= \frac{1}{F'}\left(V_v + \frac{F''}{F'}V\right)_v \\ &= \frac{1}{F'}\left(V_{vv} + \frac{F''}{F'}V_v + \left(\frac{F''}{F'}\right)' V\right) \end{aligned} \tag{3.62}$$

however from (3.61), (3.62) becomes

$$W_{ww} = \frac{1}{F'}\left(V_{vv} - \frac{1}{2}V_v\right),$$

and therefore we see that (3.55e) is connected. Therefore, to connect the determining equations (3.55), we only need to solve (3.61), which is easily solved giving

$$F = \mathrm{e}^{-v/2}$$

noting that we have suppressed the constants of integration. Therefore, the transformation connecting the determining equations is

$$w = e^{-v/2},$$

and this is precisely the same transformation that connects the equations. Further, substitution into $u = v_x$ gives rise to

$$u = -2\frac{w_x}{w},$$

the famous Hopf–Cole transformation known to link solutions of the heat equation to Burgers' equation (see Hopf [8] and Cole [9]).

3.2.4 Heat Equation with a Source

The power of symmetry analysis becomes evident when we attempt to construct exact solutions to general classes of nonlinear PDE important in modeling a wide variety of phenomena. For example, the heat equation with a source term

$$u_t = u_{xx} + F(u), \quad F'' \neq 0 \tag{3.63}$$

has a number of applications including temperature variations due to microwave heating. Here, we will perform a symmetry analysis on this equation. We will present cases for F where the equation admits special symmetries (see Dorodnitsyn [10]). We will further consider these symmetries where reduction to ODEs will be performed.

We define Δ such that

$$\Delta = u_t - u_{xx} - F(u).$$

Lie's invariance is

$$\Gamma^{(2)}\Delta\Big|_{\Delta=0} = 0.$$

This, in turn, gives

$$U_{[t]} - U_{[xx]} - F'(u)U = 0.$$

Substitution of the extension's $U_{[t]}$ and $U_{[xx]}$ gives

$$U_t + \left(U_u - T_t\right) u_t - X_t u_x - T_u u_t^2 - X_u u_t u_x$$
$$- U_{xx} - T_{xx} u_t - \left(2U_{xu} - X_{xx}\right) u_x - 2T_{xu} u_t u_x$$

$$
\begin{aligned}
&-\left(U_{uu}-2X_{xu}\right)u_x^2-T_{uu}u_tu_x^2-X_{uu}u_x^3\\
&-2T_xu_{tx}-\left(U_u-2X_x\right)u_{xx}-2T_uu_xu_{tx}\\
&-T_uu_tu_{xx}-3X_uu_xu_{xx}-F'(u)U=0.
\end{aligned}
$$

Eliminating u_{xx} where appropriate using (3.63) and isolating coefficients involving u_t, u_x and u_{tx} and various products gives rise to the following set of determining equations:

$$U_t-U_{xx}+\left(U_u-2X_x\right)F(u)-F'(u)U=0, \tag{3.64a}$$

$$T_{xx}-T_t+2X_x-F(u)T_u=0, \tag{3.64b}$$

$$X_{xx}-X_t-2U_{xu}-3X_uF(u)=0, \tag{3.64c}$$

$$T_{xu}+X_u=0, \tag{3.64d}$$

$$2X_{xu}-U_{uu}=0, \tag{3.64e}$$

$$T_{uu}=0, \tag{3.64f}$$

$$X_{uu}=0, \tag{3.64g}$$

$$T_x=0, \tag{3.64h}$$

$$T_u=0. \tag{3.64i}$$

From (3.64), we immediately find that

$$T_x=0,\qquad T_u=0,\qquad X_u=0,\qquad U_{uu}=0, \tag{3.65}$$

leaving

$$U_t-U_{xx}+\left(U_u-2X_x\right)F(u)-F'(u)U=0, \tag{3.66a}$$

$$2X_x-T_t=0, \tag{3.66b}$$

$$X_{xx}-X_t-2U_{xu}=0. \tag{3.66c}$$

From (3.65), we have that

$$T=A(t),\qquad X=B(t,x),\qquad U=P(t,x)u+Q(t,x),$$

where A, B, P and Q are arbitrary functions. With these assignments, equation (3.66) becomes

$$(Pu+Q)\,F'(u)+\left(2B_x-P\right)F(u)=\left(P_t-P_{xx}\right)u+Q_t-Q_{xx}, \tag{3.67a}$$

$$2B_x - A_t = 0, \tag{3.67b}$$

$$B_{xx} - B_t - 2P_x = 0. \tag{3.67c}$$

From (3.67a), we see four cases arise:

$$\begin{array}{lll} \text{(i)} & P = 0, & Q = 0, \\ \text{(ii)} & P = 0, & Q \neq 0, \\ \text{(iii)} & P \neq 0, & Q = 0, \\ \text{(iv)} & P \neq 0, & Q \neq 0. \end{array}$$

Case (i) $P = 0, \quad Q = 0$
In this case, equation (3.67a) becomes

$$2B_x F(u) = 0,$$

from which we deduce that $B_x = 0$ as $F \neq 0$. Further, from (3.67c), $B_t = 0$ giving that B is a constant. Finally, from (3.67b), $A_t = 0$, but since $A = A(t)$ gives that A is also a constant. Therefore, we have the following: in the case of arbitrary $F(u)$, the only symmetry that is admitted is

$$T = c_1, \quad X = c_2, \quad U = 0, \tag{3.68}$$

where here, and hereinafter c_i, $i = 1, 2, 3, \ldots$ are constant.

Case (ii) $P = 0, \quad Q \neq 0$
In this case, equation (3.67a) becomes

$$QF'(u) + 2B_x F(u) = Q_t - Q_{xx},$$

which dividing through by Q gives

$$F'(u) + \frac{2B_x}{Q} F(u) = \frac{Q_t - Q_{xx}}{Q}. \tag{3.69}$$

By differentiating (3.69) with respect to t, x and u, it is an easy matter to deduce that

$$\frac{2B_x}{Q} = m, \qquad \frac{Q_t - Q_{xx}}{Q} = k_1, \tag{3.70}$$

where m and k_1 are arbitrary constants. With these assignments, (3.69) becomes

$$F'(u) + mF(u) = k_1,$$

which is easily solved giving

$$F(u) = \frac{k_1}{m} + k_2\,\mathrm{e}^{-mu},$$

where k_2 is an additional arbitrary constant. We note that if $m = 0$, then from (3.69) $F'' = 0$, which is not of interest. From (3.67), we have

$$2B_x - A_t = 0,$$
$$B_{xx} - B_t = 0,$$

from which we can deduce $B_{xx} = 0$ so $B_t = 0$ and $A'' = 0$. This then leads to

$$A = 2c_1 t + c_0, \qquad B = c_1 x + c_2,$$

and, in turn, leads to

$$Q = \frac{2c_1}{m},$$

giving $k_1 = 0$ from (3.70). Therefore, we have the following: If

$$F(u) = k_2\,\mathrm{e}^{-mu}$$

the infinitesimal transformations T, X and U are

$$T = 2c_1 t + c_0, \qquad X = c_1 x + c_2, \qquad U = \frac{2c_1}{m}. \tag{3.71}$$

Case (iii) $P \neq 0, \quad Q = 0$
In this case, equation (3.67a) becomes

$$PuF'(u) + \left(2B_x - P\right) F(u) = \left(P_t - P_{xx}\right) u,$$

which dividing through by P gives

$$uF'(u) + \frac{2B_x - P}{P} F(u) = \frac{P_t - P_{xx}}{P} u. \tag{3.72}$$

Again, from (3.72), it is an easy matter to find that

$$\frac{2B_x - P}{P} = m, \qquad \frac{P_t - P_{xx}}{P} = k_1, \tag{3.73}$$

where m and k_1 are arbitrary constants. With these assignments, (3.72) becomes

$$uF'(u) + mF(u) = k_1 u,$$

which is easily solved giving

$$F(u) = \begin{cases} \left(k_1 \ln u + k_2\right) u, & \text{if } m = -1 \\ -\dfrac{k_1 u}{m+1} + k_2\, u^{-m}, & \text{if } m \neq -1 \end{cases},$$

where k_2 is an additional arbitrary constant. As the case $m = -1$ presents itself as special, we consider it first. From (3.73) and (3.67b), we see that

$$B_x = 0, \quad A_t = 0,$$

which must be solved in conjunction with the second of (3.73) and (3.67c). These are easily solved giving

$$A = c_1, \quad B = c_2 \mathrm{e}^{k_1 t} + c_2, \quad P = \left(-\frac{1}{2} k_1 c_2 x + c_4\right) \mathrm{e}^{k_1 t}$$

This then gives the following: If

$$F(u) = \left(k_1 \ln u + k_2\right) u,$$

the infinitesimal transformations T, X, and U are

$$T = c_1, \qquad X = c_2 e^{k_1 t} + c_3, \qquad U = \left(-\frac{1}{2} k_1 c_2 x + c_4\right) e^{k_1 t}\, u. \quad (3.74)$$

If $m \neq -1$, then from (3.73), we find

$$P = \frac{2B_x}{m+1}. \quad (3.75)$$

From (3.67c), we have

$$2B_x - A_t = 0,$$

from which we can deduce $B_{xx} = 0$ and thus from (3.75) $P_x = 0$. This then leads to (from (3.67c)) $B_t = 0$ and thus

$$A = 2c_1 t + c_0, \quad B = c_1 x + c_2, \quad P = \frac{2c_1}{m+1}$$

giving $k_1 = 0$ from (3.73). Therefore, we have the following: If

$$F(u) = k_2\, u^{-m}$$

the infinitesimal transformations T, X and U are

$$T = 2c_1 t + c_0, \qquad X = c_1 x + c_2, \qquad U = \frac{2c_1}{m+1}\, u. \tag{3.76}$$

Case (iv) $P \neq 0, \quad Q \neq 0$
In this case, equation (3.67a) becomes

$$(Pu + Q)\, F'(u) - \left(P - 2B_x\right) F(u) = \left(P_t - P_{xx}\right) u + Q_t - Q_{xx},$$

which dividing through by P gives

$$\left(u + \frac{Q}{P}\right) F'(u) + \frac{2B_x - P}{P} F(u) = \frac{P_t - P_{xx}}{P} u + \frac{Q_t - Q_{xx}}{P}. \tag{3.77}$$

From (3.77), we can deduce that

$$\frac{Q}{P} = a, \quad \frac{2B_x - P}{P} = m, \quad \frac{P_t - P_{xx}}{P} = k_1, \quad \frac{Q_t - Q_{xx}}{P} = k_2, \tag{3.78}$$

where a, m, k_1, and k_2 are arbitrary constants. With these assignments, we can deduce from (3.78) that $k_2 = ak_1$ giving (3.77) becomes

$$(u + a)F'(u) + mF(u) = k_1(u + a). \tag{3.79}$$

Comparing (3.79) and (3.2.4) shows that they are the same if we let $u + a \to u$ and thus cases (iii) and (iv) become one in the same. Therefore, the results for case (iv) are obtained by merely replacing u by $u + a$ in case (iii).

Reductions

Here, we focus our attention to a class of equation for exponential, logarithmic and power law type source terms.

EXAMPLE 3.13 Exponential Source Terms

We found that the heat equation with an exponential source term

$$u_t = u_{xx} + k_1 e^{mu} \tag{3.80}$$

led to the symmetries given in (3.71). For this example, we will choose $c_1 = 1$ with all other $c_i's$ zero. We will also choose $k_2 = 1$ and $m = -1$ as scaling of t. The invariant surface condition is

$$2tu_t + xu_x = -2,$$

which has the solution

$$u = -\ln t + f\left(\frac{x}{\sqrt{t}}\right).$$

Upon substitution into (3.80) leads to

$$-\frac{r}{2}f' - 1 = f'' + e^f,$$

where $r = x/\sqrt{t}$. This is a nonlinear ODE in f that needs to be solved. ∎

EXAMPLE 3.14 Logarithmic Source Terms

We found that the heat equation with an logarithmic source term

$$u_t = u_{xx} + \left(k_1 \ln u + k_2\right) u \tag{3.81}$$

led to the symmetries given in (3.74). For this example, we will choose $c_1 = 1$, $c_2 = -2$, $c_3 = 0$, $c_4 = 0$, $k_1 = 1$ and $k_2 = 0$. The invariant surface condition is

$$u_t - 2e^t u_x = xe^t u,$$

which has the solution

$$u = e^{-x^2/4} f\left(x + 2e^t\right). \tag{3.82}$$

Upon substitution into (3.81) leads to

$$rf' = f'' + f \ln f,$$

where $r = x + 2e^t$. ∎

EXAMPLE 3.15 Power Law Source Terms

We found that the heat equation with a power law source term

$$u_t = u_{xx} + k_2 u^{-m} \tag{3.83}$$

led to the symmetries given in (3.76). For this example, we will choose $c_1 = 1$ with all other $c_i' s$ zero. We will also choose $k_2 = 1$. The invariant surface condition is

$$2tu_t + xu_x = \frac{2u}{m+1},$$

which has the solution

$$u = t^{\frac{1}{m+1}} f\left(\frac{x}{\sqrt{t}}\right).$$

Upon substitution into (3.83) leads to

$$\frac{1}{m+1} f - \frac{r}{2} f' = f'' + f^{-m},$$

where $r = x/\sqrt{t}$. ∎

EXAMPLE 3.16 Arbitrary Source Terms

We found that the heat equation with an arbitrary source term led to the symmetries given in (3.68). The invariant surface condition is

$$c_1 u_t + c_2 u_x = 0,$$

and if we set $c_2 = c_1 c$, then this becomes

$$u_t + cu_x = 0,$$

which has the solution

$$u = f(x - ct).$$

Upon substitution into (3.63) leads to

$$-cf' = f'' + F(f).$$

∎

EXERCISES

1. Calculate the symmetries for the the Folker–Planck equation

$$u_t = u_{xx} + (xu)_x.$$

2. Calculate the symmetries for the the nonlinear diffusion equation

$$u_t = \frac{u_{xx}}{u_x^2}.$$

Compare these with the symmetries of the heat equation and derive a transformation that linearizes this equation.

3. Calculate the symmetries for the the nonlinear diffusion equation and its potential form

$$u_t = \left(\frac{u_x}{u^2+1}\right)_x, \quad u_t = \frac{u_{xx}}{u_x^2+1}.$$

and compare their symmetries.

4. Calculate the symmetries for the the nonlinear wave equation

$$u_{tt} = uu_{xx}.$$

Use a particular symmetry to reduce the PDE to an ODE.

5. Calculate the symmetries for the the Fisher's equation

$$u_t = u_{xx} + u(1-u).$$

Use a particular symmetry to reduce the PDE to an ODE.

6. Calculate the symmetries for the the Fitzhugh–Nagumo equation

$$u_t = u_{xx} + u(1-u)(u-a).$$

Use a particular symmetry to reduce the PDE to an ODE.

7. Calculate the symmetries of the minimal surface equation

$$\left(1+u_y^2\right)u_{xx} - 2u_x u_y u_{xy} + \left(1+u_x^2\right)u_{yy} = 0.$$

Use a particular symmetry to reduce the PDE to an ODE.

*8. Classify the symmetries of the Schrodinger equation

$$u_t = u_{xx} + V(x)u.$$

*9. Classify the symmetries of the nonlinear diffusion equation

$$u_t = \left(D(u)u_x\right)_x,$$

(see Ovsjannikov [11] and Bluman [12]).

*10. Classify the symmetries of the nonlinear diffusion–convection equation

$$u_t = \left(D(u)u_x\right)_x - K'(u)u_x,$$

(see Edwards [13]).

*11. Classify the symmetries of nonlinear wave equation

$$u_{tt} = \left(c^2(u)u_x\right)_x,$$

and the linear wave equation

$$u_{tt} = c^2(x)u_{xx}$$

(see Ames *et al.* [14] and Bluman and Kumei [5]).

3.3 HIGHER ORDER PDEs

Previously, we constructed the symmetries of second-order PDEs. We now extend symmetries to higher order equations. Consider the general nth order PDE

$$\Delta\left(t, x, u, u_t, u_x, u_{tt}, u_{tx}, u_{xx}, \ldots, u_{t(n)}, \cdots, u_{t(n-i)x(i)}, \cdots u_{x(n)}\right) = 0,$$

where for convenience, we have denoted $u_{t(n-i)x(i)} = \partial_t^{n-i}\partial_x^i u$. We again introduce the infinitesimal operator Γ as

$$\Gamma = T\frac{\partial}{\partial t} + X\frac{\partial}{\partial x} + U\frac{\partial}{\partial u},$$

where $T = T(t, x, u), X = X(t, x, u)$ and $U = U(t, x, u)$ are to be determined. We define the nth extension to the operator Γ as $\Gamma^{(n)}$, given recursively by

$$\Gamma^{(n)} = \Gamma^{(n-1)} + \sum_{i=0}^{n} U_{[t(n-i)x(i)]}\frac{\partial^n}{\partial t^{n-i}\partial x^i}.$$

Lie's invariance condition becomes

$$\Gamma^{(n)}\Delta\Big|_{\Delta=0} = 0, \tag{3.84}$$

where the extended transformations are given as

$$\begin{aligned}
U_{[t]} &= D_t(U) - u_t D_t(T) - u_x D_t(X),\\
U_{[x]} &= D_x(U) - u_t D_x(T) - u_x D_x(X),\\
&\vdots\\
U_{[t(n-i)x(i)]} &= D_t(U_{[t(n-i-1)x(i)]}) - u_{[t(n-i)x(i)]}D_t(T)\\
&\quad - u_{[t(n-i-1)x(i+1)]}D_t(X) \text{ or,}
\end{aligned}$$

$$= D_x(U_{[t(n-i)x(i-1)]}) - u_{[t(n-i+1)x(i-1)]}D_x(T)$$
$$-u_{[t(n-i)x(i)]}D_x(X).$$

The total derivative operators are given by

$$D_t = \frac{\partial}{\partial t} + u_t\frac{\partial}{\partial u} + \cdots + u_{t(i+1)x(j)}\frac{\partial^{i+j}}{\partial u_{t(i)x(j)}},$$
$$D_x = \frac{\partial}{\partial x} + u_x\frac{\partial}{\partial u} + \cdots + u_{t(i)x(j+1)}\frac{\partial^{i+j}}{\partial u_{t(i)x(j)}}.$$

For example, $U_{[ttx]}$ is given by

$$U_{[ttx]} = D_t(U_{[tx]}) - u_{ttx}D_t(T) - u_{txx}D_t(X), \quad \text{or}$$
$$= D_x(U_{[tt]}) - u_{ttt}D_x(T) - u_{ttx}D_x(X).$$

EXAMPLE 3.17 Korteweg–DeVries Equation

To illustrate, we consider the KdV (Korteweg-de Vries equation

$$u_t + uu_x + u_{xxx} = 0. \tag{3.85}$$

Expanding Lie's invariance condition (3.84) gives

$$U_{[t]} + uU_{[x]} + Uu_x + U_{[xxx]} = 0.$$

In this case, we will need the first extensions $U_{[t]}$, $U_{[x]}$ and the third extension $U_{[xxx]}$, namely

$$U_{[t]} = D_t(U) - u_tD_t(T) - u_xD_t(X),$$
$$U_{[x]} = D_x(U) - u_tD_x(T) - u_xD_x(X),$$
$$U_{[xxx]} = D_x(U_{[xx]}) - u_{txx}D_x(T) - u_{xxx}D_x(X),$$

noting that $U_{[xx]}$ will be needed. As we have seen previously, we expand Lie's invariance condition and substitute the original equation, $u_{xxx} = -u_t - uu_x$ and isolate coefficients involving u_t, u_x, u_{tx}, u_{xx}, u_{ttt}, u_{ttx} and u_{txx} and various products. This gives

$$U_t + uU_x + U_{xxx} = 0, \tag{3.86a}$$
$$T_t + uT_x - 3X_x + T_{xxx} = 0, \tag{3.86b}$$
$$X_t - 2uX_x + X_{xxx} - 3U_{xxu} - U = 0, \tag{3.86c}$$

$$X_u - T_{xxu} = 0, \tag{3.86d}$$

$$uX_u - X_{xxu} + U_{xuu} = 0, \tag{3.86e}$$

$$T_{xuu} = 0, \quad U_{uuu} - 3X_{xuu} = 0, \quad T_{uuu} = 0, \tag{3.86f}$$

$$X_{uuu} = 0, \quad T_{xx} = 0, \quad U_{xu} - X_{xx} = 0, \tag{3.86g}$$

$$T_{xu} = 0, \quad U_{uu} - X_{xu} = 0, \quad X_{uu} = 0, \tag{3.86h}$$

$$T_x = 0, \quad T_u = 0. \tag{3.86i}$$

From (3.86), we have

$$T_x = 0, \quad T_u = 0, \quad X_u = 0, \quad U_{uu} = 0,$$

from which it follows that

$$T = A(t), \quad X = B(t, x), \quad U = P(t, x)u + Q(t, x),$$

where $A = A(t)$, $B = B(t, x)$, $P = P(t, x)$ and $Q = Q(t, x)$ are arbitrary functions. Thus, from (3.86), we have the following

$$\left(P_t u + Q_t\right) + \left(P_x u + Q_x\right) u + P_{xxx} u + Q_{xxx} = 0, \tag{3.87a}$$

$$B_t - 2uB_x + B_{xxx} - 3P_{xx} - Pu - Q = 0, \tag{3.87b}$$

$$A_t - 3B_x = 0, \tag{3.87c}$$

$$P_x - B_{xx} = 0. \tag{3.87d}$$

Isolating the coefficients of u in (3.87a) and (3.87b) gives the following final set of determining equations to solve:

$$A_t - 3B_x = 0, \tag{3.88a}$$

$$P_x - B_{xx} = 0, \tag{3.88b}$$

$$P + 2B_x = 0, \tag{3.88c}$$

$$P_x = 0, \tag{3.88d}$$

$$Q_t + Q_{xxx} = 0, \tag{3.88e}$$

$$P_t + Q_x + P_{xxx} = 0, \tag{3.88f}$$

$$3P_{xx} - B_t - B_{xx} + Q = 0. \tag{3.88g}$$

Solving (3.88) gives

$$T = 3c_1 t + c_0, \quad X = c_3 t + c_1 x + c_2, \quad U = -2c_1 u + c3. \tag{3.89}$$

We now use these to obtain a symmetry reduction of the original PDE. We will consider two examples. ■

EXAMPLE 3.17a Symmetry Reduction 1

If we set

$$c_0 = 1, \quad c_1 = 0, \quad c_2 = c, \quad c_3 = 0,$$

in (3.89) where c is an arbitrary constant, then we obtain the invariant surface condition

$$u_t + cu_x = 0.$$

By the method of characteristics, we obtain the solution as

$$u = f(x - ct).$$

Substitution into the original equation (3.85) gives rise to the ODE

$$f''' + ff' - cf' = 0,$$

where the argument r is defined as $r = x - ct$. Integrating once and suppressing the constant of integration gives

$$f'' + \frac{1}{2}f^2 - cf = 0. \tag{3.90}$$

One particular solution of (3.90) is

$$f = 12p^2 \mathrm{sech} px,$$

where the constant c is chosen as $c = 4p^2$. This solution is commonly known as the "one soliton" solution. ■

EXAMPLE 3.17b Symmetry Reduction 2

If we set

$$c_0 = 0, \quad c_1 = 1, \quad c_2 = 0, \quad c_3 = 0,$$

in (3.89), then we obtain the invariant surface condition

$$3tu_t + xu_x = -2u.$$

By the method of characteristics, we obtain the solution as

$$u = t^{-2/3} f\left(\frac{x}{t^{1/3}}\right).$$

Substitution into the original equation (3.85) gives rise to the ODE

$$f''' + ff' - \frac{1}{3}rf' - \frac{2}{3}f = 0,$$

where $r = x/t^{\frac{1}{3}}$. Solutions of this would then lead to exact solution of the KdV equation. ■

EXAMPLE 3.18

Here we calculate the symmetries of the Boussinesq equation

$$u_{tt} + uu_{xx} + u_x^2 + u_{xxxx} = 0. \tag{3.91}$$

Expanding Lie's invariance condition gives

$$U_{[tt]} + Uu_{xx} + uU_{[xx]} + 2u_x U_{[x]} + U_{[xxxx]} = 0. \tag{3.92}$$

In this case. we will need the extensions $U_{[x]}$, $U_{[tt]}$, $U_{[xx]}$ and $U_{[xxxx]}$ given by

$$\begin{aligned}
U_{[x]} &= D_x(U) - u_t D_x(T) - u_x D_x(X), \\
U_{[tt]} &= D_t(U_{[t]}) - u_{tt} D_t(T) - u_{tx} D_t(X), \\
U_{[xx]} &= D_x(U_{[x]}) - u_{tx} D_x(T) - u_{xx} D_x(X), \\
U_{[xxxx]} &= D_x(U_{[xxx]}) - u_{txxx} D_x(T) - u_{xxxx} D_x(X).
\end{aligned}$$

Expanding (3.92) and imposing the original equation $u_{xxxx} = -u_{tt} - uu_{xx} - u_x^2$ gives on isolating the coefficients of the derivatives of u

$$T_x = 0, \quad T_u = 0, \tag{3.93a}$$

$$X_t = 0, \quad X_u = 0, \tag{3.93b}$$

$$U_{uu} = 0, \tag{3.93c}$$

$$2X_x - T_t = 0, \tag{3.93d}$$

$$U_u + 2X_x = 0, \tag{3.93e}$$

$$2U_{tu} - T_{tt} = 0, \tag{3.93f}$$

$$2U_{xu} - 3X_{xx} = 0, \tag{3.93g}$$

$$U + 6U_{xxu} + 2uX_x - 4X_{xxx} = 0, \tag{3.93h}$$

$$4U_{xxxu} + 2U_x + 2uU_{xu} - X_{xxxx} - uX_{xx} = 0, \tag{3.93i}$$

$$U_{tt} + uU_{xx} + U_{xxxx} = 0. \tag{3.93j}$$

From (3.93a), (3.93b) and (3.93c) give

$$T = T(t), \quad X = X(x), \quad U = P(t, x)u + Q(t, x).$$

Further, differentiating (3.93d) with respect to t and x gives

$$T_{tt} = 0, \quad X_{xx} = 0,$$

which leads to, in conjunction with (3.93d)

$$T = 2c_1 t + c_2, \quad X = c_1 x + c_3.$$

From (3.93f) and (3.93g), we find that

$$U_{tu} = 0, \quad U_{xu} = 0,$$

which from (3.93e) gives

$$U = -2c_1 u + Q(t, x).$$

Finally, substitution into the remaining equations of (3.93) gives $Q = 0$. Thus, the infinitesimals are

$$T = 2c_1 t + c_2, \quad X = c_1 x + c_3, \quad U = -2c_1 u. \tag{3.94}$$

As an exemplary reduction ,we set $c_1 = 1$ and $c_2 = c_3 = 0$ in (3.94). The associated invariant surface condition is

$$2tu_t + xu_x = -2u,$$

noting that we can cancel the arbitrary constant c_1. Solving this gives

$$u = \frac{1}{t} f\left(\frac{x}{\sqrt{t}}\right)$$

and substitution into (3.93) gives

$$f^{(4)} + \left(f + \frac{r^2}{4}\right) f'' + f'^2 + \frac{7}{4} r f' + 2f = 0,$$

where $r = xt^{-1/2}$. ■

EXERCISES

1. Calculate the symmetries for the modified KdV and potential KdV and Harry–Dym equations

$$u_t - u^2 u_x + u_{xxx} = 0,$$
$$u_t + u_x^2 + u_{xxx} = 0,$$
$$u_t - u^3 u_{xxx} = 0,$$

2. Calculate the symmetries for the following appearing in Boundary Layer theory

$$u_x u_{yy} - u_y u_{xy} = u_{yyy}. \tag{3.95}$$

3. Calculate the symmetries for the thin film equation

$$h_t = \left(h^n h_{xxx}\right)_x. \tag{3.96}$$

(Gandarias and Medina [15])

4. Calculate the symmetries for the following appearing in the growth of grain boundaries

$$u_t = \left[u^{-1}\left(u^{-3} u_x\right)_x\right]_{xx} \tag{3.97}$$

(Broadbridge and Tritscher [16]).

3.4 SYSTEMS OF PDEs

In this section, we consider the symmetries of systems of partial differential equations. In particular, we consider a system of two equations and two independent variables but the analysis easily extends to more equations and more independent variables.

3.4.1 First-Order Systems

Consider

$$F\left(t, x, u, v, u_t, v_t, u_x, v_x\right) = 0, \tag{3.98a}$$
$$G\left(t, x, u, v, u_t, v_t, u_x, v_x\right) = 0. \tag{3.98b}$$

Invariance of (3.98) is conveniently written as

$$\Gamma^{(1)}F\Big|_{F=0,G=0} = 0, \qquad \Gamma^{(1)}G\Big|_{F=0,G=0} = 0,$$

where Γ is defined as

$$\Gamma = T\frac{\partial}{\partial t} + X\frac{\partial}{\partial x} + U\frac{\partial}{\partial U} + V\frac{\partial}{\partial V} \tag{3.99}$$

and $\Gamma^{(1)}$ is the extension to the operator Γ in (3.99), namely

$$\Gamma^{(1)} = \Gamma + U_{[t]}\frac{\partial}{\partial u_t} + U_{[x]}\frac{\partial}{\partial u_x} + V_{[t]}\frac{\partial}{\partial v_t} + V_{[x]}\frac{\partial}{\partial v_x}.$$

The extended transformations are given by

$$U_{[t]} = D_t(U) - u_t D_t(T) - u_x D_t(X), \tag{3.100a}$$

$$U_{[x]} = D_x(U) - u_t D_x(T) - u_x D_x(X), \tag{3.100b}$$

$$V_{[t]} = D_t(V) - v_t D_t(T) - v_x D_t(X), \tag{3.100c}$$

$$V_{[x]} = D_x(V) - v_t D_x(T) - v_x D_x(X). \tag{3.100d}$$

The total differential operators D_t and D_x are given, respectively, by

$$D_t = \frac{\partial}{\partial t} + u_t\frac{\partial}{\partial u} + u_{tt}\frac{\partial}{\partial u_t} + u_{tx}\frac{\partial}{\partial u_x} + \cdots + v_t\frac{\partial}{\partial v} + v_{tt}\frac{\partial}{\partial v_t} + v_{tx}\frac{\partial}{\partial v_x} + \cdots,$$

$$D_x = \frac{\partial}{\partial x} + u_x\frac{\partial}{\partial u} + u_{tx}\frac{\partial}{\partial u_t} + u_{xx}\frac{\partial}{\partial u_x} + \cdots + v_x\frac{\partial}{\partial v} + v_{tx}\frac{\partial}{\partial v_t} + v_{xx}\frac{\partial}{\partial v_x} + \cdots.$$

Once the infinitesimals T, X, U, and V have been found, the associated invariant surface conditions are

$$Tu_t + Xu_x = U, \quad Tv_t + Xv_x = V. \tag{3.101}$$

EXAMPLE 3.19

Here we calculate the symmetries of the system of equations

$$v_x = u, \qquad v_t = \frac{u_x}{1+u^2}, \tag{3.102}$$

that is equivalent to the heat equation

$$u_t = \left(\frac{u_x}{1+u^2}\right)_x.$$

Lie's invariance condition for (3.102) is

$$V_{[x]} = U, \qquad V_{[t]} = \frac{U_{[x]}}{1+u^2} - \frac{2u\,U}{(1+u^2)^2}u_x. \tag{3.103}$$

The extensions $U_{[x]}$, $V_{[t]}$ and $V_{[x]}$ are given in (3.100). Expanding (3.103) and imposing the original system equation (3.102) gives on isolating the coefficients of u_t and v_t

$$T_u = 0, \tag{3.104a}$$

$$(1+u^2)V_u - u(1+u^2)X_u - T_x - uT_v = 0 \tag{3.104b}$$

$$V_x + (V_v - X_x)u - u^2X_v - U = 0, \tag{3.104c}$$

$$(1+u^2)X_u - T_v = 0, \tag{3.104d}$$

$$(1+u^2)V_t - U_x - uU_v - u(1+u^2)X_t = 0, \tag{3.104e}$$

$$T_x + uT_v - u(1+u^2)X_u + (1+u^2)V_u = 0, \tag{3.104f}$$

$$(1+u^2)(V_v - U_u + X_x - T_t) + 2uU = 0. \tag{3.104g}$$

Adding and subtracting (3.104b) and (3.104f) gives

$$V_u - uX_u = 0, \quad T_x + uT_v = 0. \tag{3.105}$$

From (3.105) and (3.104a) give that

$$T = T(t).$$

From (3.104d) gives that $X_u = 0$ which further from (3.105) gives $V_u = 0$. From (3.104c), we solve for U giving

$$U = V_x + (V_v - X_x)u - X_vu^2. \tag{3.106}$$

As T, X and V are independent of u and U is quadratic is u, substitution of (3.106) into (3.104c) and (3.104g) and isolating coefficients of u gives

$$2X_x - T_t = 0, \tag{3.107a}$$

$$2V_v - T_t = 0, \tag{3.107b}$$

$$X_v + V_x = 0, \tag{3.107c}$$

$$V_t - V_{xx} = 0, \tag{3.107d}$$

$$X_t - X_{vv} = 0, \tag{3.107e}$$

$$X_t - X_{xx} + 2V_{xv} = 0, \tag{3.107f}$$

$$V_t - V_{vv} + 2X_{xv} = 0. \tag{3.107g}$$

Differentiating (3.107a) and (3.107b) with respect to t and v, we have

$$X_{xx} = 0, \quad X_{xv} = 0, \quad V_{xv} = 0, \quad V_{vv} = 0.$$

Thus from (3.107f) and (3.107g)

$$X_t = 0, \quad V_t = 0.$$

From either (3.107a) or (3.107b) we deduce that

$$T_{tt} = 0 \quad \Rightarrow \quad T = 2c_1 t + c_0, \tag{3.108}$$

noting that the factor of 2 becomes apparent in a moment. From (3.107a) and (3.107b) with T given in (3.108) we have

$$2X_x - 2c_1 = 0, \quad 2V_v - 2c_1 = 0,$$

which integrates yielding

$$X = c_1 x + A(v), \quad V = c_1 v + B(x), \tag{3.109}$$

where A and B are arbitrary functions of integration. Substituting (3.109) into (3.107d) and (3.107e) gives

$$A''(v) = 0, \quad B''(x) = 0,$$

leading to (subject to (3.107c))

$$A(v) = c_2 v + c_3, \quad B(x) = -c_2 x + c_4.$$

The entire system (3.104) has now been solved leading to

$$T = 2c_1 t + c_0, \quad X = c_1 x + c_2 v + c_3,$$
$$U = -c_2(1 + u^2), \quad V = -c_2 x + c_1 v + c_4. \tag{3.110}$$

We now consider two reductions associated with the constants c_1 and c_2. ■

EXAMPLE 3.19a Reduction 1

The invariant surface condition (3.101) associated with c_1 in (3.110) is

$$2tu_t + xu_x = 0, \quad 2tv_t + xv_x = v.$$

The solution of these is

$$u = f\left(\frac{x}{\sqrt{t}}\right), \qquad v = \sqrt{t}g\left(\frac{x}{\sqrt{t}}\right),$$

and substitution into the original system (3.102) gives

$$g' = f, \quad \frac{1}{2}(g - rg') = \frac{f'}{1 + f^2},$$

where $r = x/\sqrt{t}$. ■

EXAMPLE 3.19b Reduction 2

The invariant surface condition (3.101) associated with $c_0 = 1$ and $c_2 = 1$ with the rest zero in (3.110) is

$$u_t + vu_x = -(1 + u^2), \quad v_t + vv_x = -x. \tag{3.111}$$

The solution of the second is

$$\tan^{-1}\left(\frac{v}{x}\right) + t = F\left(x^2 + v^2\right). \tag{3.112}$$

As it is impossible to isolate v to solve the first invariant surface condition in (3.111), we proceed in a different direction. From the original system (3.102), we have $v_x = u$, then upon elimination of u in the original equation (3.102), we obtain

$$v_t = \frac{v_{xx}}{1 + v_x^2} \tag{3.113}$$

and substitution of (3.112) into (3.113) gives

$$4rF'' + 8r^2F'^{\,3} + 4r^2F'^{\,2} + 6F' + 1 = 0,$$

which is clearly difficult. Therefore, we will take an alternate route. Eliminating v_t from (3.111) and (3.102) and u_x gives

$$\frac{u_x}{1+u^2} + vv_x + x = 0, \quad \frac{u_t}{1+u^2} + vv_t + 1 = 0. \tag{3.114}$$

Integrating each equation, respectively, in (3.114) gives

$$\tan^{-1} u + \frac{1}{2}(v^2 + x^2) + f(t) = 0, \quad \tan^{-1} u + \frac{1}{2}v^2 + t + g(x) = 0,$$

where $f(t)$ and $g(x)$ are arbitrary functions of integration. As these must be the same give the final solution as

$$\tan^{-1} u + \frac{1}{2}(v^2 + x^2) + t = c, \tag{3.115}$$

noting that we can set the arbitrary constant c to zero without the loss of generality. Replacing $u = v_x$ in (3.115) and using (3.112) gives

$$F' = -\frac{1}{2r}\tan\left(F + \frac{r}{2}\right),$$

where $r = x^2 + v^2$. ■

3.4.2 Second-Order Systems

Consider

$$F\left(x, y, u, v, u_x, v_x, u_y, v_y, u_{xx}, \ldots v_{yy}\right) = 0, \tag{3.116a}$$

$$G\left(x, y, u, v, u_x, v_x, u_y, v_y, u_{xx}, \ldots v_{yy}\right) = 0. \tag{3.116b}$$

Invariance of (3.116) is conveniently written as

$$\Gamma^{(2)}F\Big|_{F=0,G=0} = 0, \quad \Gamma^{(2)}G\Big|_{F=0,G=0} = 0,$$

where Γ is defined as

$$\Gamma = X\frac{\partial}{\partial x} + Y\frac{\partial}{\partial y} + U\frac{\partial}{\partial U} + V\frac{\partial}{\partial V}, \tag{3.117}$$

and $\Gamma^{(1)}$ and $\Gamma^{(2)}$ are extensions to the operator Γ in (3.117), namely

$$\Gamma^{(1)} = \Gamma + U_{[x]}\frac{\partial}{\partial u_x} + U_{[y]}\frac{\partial}{\partial u_y} + V_{[x]}\frac{\partial}{\partial v_x} + V_{[y]}\frac{\partial}{\partial v_y},$$

$$\Gamma^{(2)} = \Gamma^{(1)} + U_{[xx]}\frac{\partial}{\partial u_{xx}} + U_{[xy]}\frac{\partial}{\partial u_{xy}} + U_{[yy]}\frac{\partial}{\partial u_{yy}}$$

$$+ V_{[xx]}\frac{\partial}{\partial v_{xx}} + V_{[xy]}\frac{\partial}{\partial v_{xy}} + V_{[yy]}\frac{\partial}{\partial v_{yy}}.$$

The extended transformations are given by

$$U_{[x]} = D_x(U) - u_x D_x(X) - u_y D_x(Y), \tag{3.118a}$$

$$U_{[y]} = D_y(U) - u_x D_y(X) - u_y D_y(Y), \tag{3.118b}$$

$$V_{[x]} = D_x(V) - v_x D_x(X) - v_y D_x(Y), \tag{3.118c}$$

$$V_{[y]} = D_y(V) - v_x D_y(X) - v_y D_y(Y), \tag{3.118d}$$

and

$$U_{[xx]} = D_x(U_{[x]}) - u_{xx} D_x(X) - u_{xy} D_x(Y), \tag{3.119a}$$

$$U_{[xy]} = D_x(U_{[y]}) - u_{xy} D_x(X) - u_{yy} D_x(Y), \tag{3.119b}$$

$$U_{[yy]} = D_y(U_{[y]}) - u_{xy} D_y(X) - u_{yy} D_y(Y), \tag{3.119c}$$

$$V_{[xx]} = D_x(V_{[x]}) - v_{xx} D_x(X) - v_{xy} D_x(Y), \tag{3.119d}$$

$$V_{[xy]} = D_x(V_{[y]}) - v_{xy} D_x(X) - v_{yy} D_x(Y), \tag{3.119e}$$

$$V_{[yy]} = D_y(V_{[y]}) - v_{xy} D_y(X) - v_{yy} D_y(Y). \tag{3.119f}$$

The total differential operators D_x and D_y are given by

$$D_x = \frac{\partial}{\partial x} + u_x\frac{\partial}{\partial u} + u_{xx}\frac{\partial}{\partial u_x} + u_{xy}\frac{\partial}{\partial u_y} + \cdots + v_x\frac{\partial}{\partial v} + v_{xx}\frac{\partial}{\partial v_x}$$

$$+ v_{xy}\frac{\partial}{\partial v_y} + \cdots$$

$$D_y = \frac{\partial}{\partial y} + u_y\frac{\partial}{\partial u} + u_{xy}\frac{\partial}{\partial u_x} + u_{yy}\frac{\partial}{\partial u_y} + \cdots + v_y\frac{\partial}{\partial v} + v_{xy}\frac{\partial}{\partial v_x}$$

$$+ v_{yy}\frac{\partial}{\partial v_y} + \cdots.$$

EXAMPLE 3.20

Consider the boundary layer equations from fluid mechanics

$$u_x + v_y = 0, \tag{3.120a}$$

$$uu_x + vu_y = u_{yy}. \tag{3.120b}$$

The invariance condition for each is given by

$$U_{[x]} + V_{[y]} = 0, \tag{3.121a}$$

$$uU_{[x]} + vU_{[y]} + Uu_x + Vu_y = U_{[yy]}. \tag{3.121b}$$

Substitution of the extended transformations (3.118) and (3.119) into both invariance conditions (3.121) gives subject first to both original equations being satisfied (plus differential consequences) the following set of determining equations:

$$U_x + V_y = 0, \quad U_v - X_y = 0, \tag{3.122a}$$

$$X_u + Y_v = 0, \quad V_u - Y_x = 0 \tag{3.122b}$$

$$U_u - V_v + Y_y - X_x = 0, \tag{3.122c}$$

and

$$X_v = 0, \quad Y_v = 0, \quad U_v = 0, \quad Y_{uu} = 0, \tag{3.123a}$$

$$2X_u - Y_v = 0, \quad U_v + 2X_y = 0, \quad X_{uu} - 2Y_{uv} = 0, \tag{3.123b}$$

$$uU_x + vU_y - U_{yy} = 0 \quad 2vY_u + 2Y_{yu} - U_{uu} = 0, \tag{3.123c}$$

$$-2uY_v + vX_v - U_{vv} - 2X_{yv} = 0, \tag{3.123d}$$

$$Y_{yy} - 2U_{yu} - uY_x + vY_y + V = 0, \tag{3.123e}$$

$$2uY_u + 2X_{yu} + 2U_{uv} - 2Y_{yv} - vY_v = 0, \tag{3.123f}$$

$$X_{yy} - vX_y - vU_v - uX_x + U + 2uY_y + 2U_{yv} = 0. \tag{3.123g}$$

From (3.123a), (3.123b), and (3.123f), we have

$$X_u = 0, \quad X_y = 0, \quad Y_u = 0,$$

which reduces (3.123) to

$$uU_x + vU_y - U_{yy} = 0, \tag{3.124a}$$

$$-uY_x - 2U_{yu} + Y_{yy} + V + vY_y = 0, \tag{3.124b}$$

$$U + 2uY_y - uX_x = 0. \qquad (3.124c)$$

From (3.124b) and (3.124c), we obtain

$$U = \left(X_x - 2Y_y\right)u, \qquad V = uY_x - vY_y - 5Y_{yy}.$$

From (3.122a) and (3.124a)

$$X_{xx} = 0, \quad Y_{xy} = 0, \quad Y_{yy} = 0,$$

thus leading to

$$X = c_1x + c_2, \qquad Y = c_3y + a(x), \qquad (3.125)$$

where $a(x)$ is an arbitrary function. This in turn gives

$$U = \left(c_1 - 2c_3\right)u, \qquad V = a'(x)u - c_3v, \qquad (3.126)$$

■

EXAMPLE 3.20a A Reduction

We now consider a reduction of the boundary layer system (3.120). However, we will proceed in a slightly different direction. If we introduce a potential $w = w(x, y)$ such that

$$u = w_y, \quad v = -w_x, \qquad (3.127)$$

then (3.120a) is automatically satisfied, whereas (3.120b) becomes

$$w_yw_{xy} - w_xw_{yy} = w_{yyy}. \qquad (3.128)$$

The associated invariant surface conditions associated with (3.120) (using (3.125) and (3.126) are

$$\left(c_1x + c_2\right)u_x + \left(c_3y + a(x)\right)u_y = \left(c_1 - 2c_3\right)u,$$
$$\left(c_1x + c_2\right)v_x + \left(c_3y + a(x)\right)v_y = a'(x)u - c_3v,$$

which becomes

$$\left(c_1x + c_2\right)w_{xy} + \left(c_3y + a(x)\right)w_{yy} = \left(c_1 - 2c_3\right)w_y, \qquad (3.129)$$
$$-\left(c_1x + c_2\right)w_{xx} - \left(c_3y + a(x)\right)w_{xy} = a'(x)w_y + c_3w_x,$$

on using (3.127). Integrating (3.129) gives

$$\left(c_1 x + c_2\right) w_x + \left(c_3 y + a(x)\right) w_y = \left(c_1 - c_3\right) w + c_4, \tag{3.130}$$

where c_4 is a constant of integration. We consider two cases. If $c_1 = 0$ and $c_2 \neq 0$, then (3.130) integrating to give

$$w = e^{mx} F\left(e^{mx} y + b(x)\right), \tag{3.131}$$

where $c_3/c_2 = -m$ and $b'(x) = -a(x)e^{mx}/c_2$. Substituting (3.131) into (3.128) gives

$$F''' + mFF'' - 2mF'^2 = 0,$$

where $F = F(r)$ and $r = e^{mx} y + b(x)$. If $c_1 \neq 0$, we can set $c_2 = 0$ without the loss of generality. The invariant surface condition (3.130) integrating to give

$$w = x^m F\left(x^m y + b(x),\right) \tag{3.132}$$

where $c_3/c_1 = -m$ and $b'(x) = -x^m a(x)/c_1 x$. Substituting (3.132) into (3.128) gives

$$F''' + (m+1)FF'' - (2m+1)F'^2 = 0,$$

where $F = F(r)$ and $r = x^m y + b(x)$. ■

EXERCISES

1. The nonlinear cubic Schrodinger equation is

$$i\psi_t + \psi_{xx} + k\psi|\psi|^2 = 0$$

or if $\psi = u + iv$

$$u_t + v_{xx} + kv\left(u^2 + v^2\right) = 0,$$
$$v_t - u_{xx} - ku\left(u^2 + v^2\right) = 0.$$

Calculate the symmetries of this system.

2. Calculate the symmetries for the following system equivalent to Burgers equation

$$v_x = 2u, \quad v_t = 2u_x - u^2$$

(Vinogradov and Krasil'shchik [17]).

3*. Calculate the symmetries for the nonlinear diffusion system

$$v_x = u, \quad v_t = K(u)u_x$$

(Bluman and Kumei [1]).

4*. Calculate the symmetries for the nonlinear wave equation system

$$u_t = v_x, \quad v_t = f(u)u_x.$$

5. Calculate the symmetries for the one-dimensional unsteady gasdynamics equations

$$\rho_t + (\rho u) = 0,$$
$$u_t + uu_x + \frac{P_x}{\rho} = 0,$$
$$P = P(\rho)$$

(see Cantwell [18] and the references within).

6. Calculate the symmetries for the two-dimensional steady Navier–Stokes equations

$$u_x + v_y = 0,$$
$$uu_x + vu_y = -\frac{p_x}{\rho} + u_{xx} + u_{yy},$$
$$uv_x + vv_y = -\frac{p_y}{\rho} + v_{xx} + v_{yy}.$$

and compare these with the unsteady version

$$u_x + v_y = 0,$$
$$u_t + uu_x + vu_y = -\frac{p_x}{\rho} + u_{xx} + u_{yy},$$
$$v_t + uv_x + vv_y = -\frac{p_y}{\rho} + v_{xx} + v_{yy}.$$

(see Cantwell [18] and the references within)

7. Calculate the symmetries for the two-dimensional steady boundary layer equations

$$u_x + v_y = 0,$$
$$u_t + uu_x + vu_y = U_t + UU_x + u_{yy},$$

where $U = U(t, x)$ is to be determined (Ma and Hui [19]).

3.5 HIGHER DIMENSIONAL PDEs

We now consider the symmetries of higher dimensional PDEs. In particular, we will consider second-order equations in three independent variables but the analysis is not restricted to only these. Consider

$$F\left(t, x, y, u, u_t, u_x, u_y, u_{tt}, u_{tx}, u_{ty} \cdots u_{yy}\right) = 0. \tag{3.133}$$

Invariance of (3.133) is conveniently written as

$$\Gamma^{(2)} F\Big|_{F=0} = 0,$$

where Γ is defined as

$$\Gamma = T\frac{\partial}{\partial t} + X\frac{\partial}{\partial x} + Y\frac{\partial}{\partial y} + U\frac{\partial}{\partial U}, \tag{3.134}$$

and $\Gamma^{(1)}$ and $\Gamma^{(2)}$ are extensions to the operator Γ in (3.134), namely

$$\Gamma^{(1)} = \Gamma + U_{[t]}\frac{\partial}{\partial u_t} + U_{[x]}\frac{\partial}{\partial u_x} + U_{[y]}\frac{\partial}{\partial u_y},$$

$$\Gamma^{(2)} = \Gamma^{(1)} + U_{[tt]}\frac{\partial}{\partial u_{tt}} + U_{[tx]}\frac{\partial}{\partial u_{tx}} + U_{[ty]}\frac{\partial}{\partial u_{ty}}$$
$$+ U_{[xx]}\frac{\partial}{\partial u_{xx}} + U_{[xy]}\frac{\partial}{\partial u_{xy}} + U_{[yy]}\frac{\partial}{\partial u_{yy}}.$$

The extended transformations are given by

$$U_{[t]} = D_t(U) - u_t D_t(T) - u_x D_t(X) - u_y D_t(Y), \tag{3.135a}$$
$$U_{[x]} = D_x(U) - u_t D_x(T) - u_x D_x(X) - u_y D_x(Y), \tag{3.135b}$$
$$U_{[y]} = D_y(U) - u_t D_y(T) - u_x D_y(X) - u_y D_y(Y), \tag{3.135c}$$

and

$$U_{[tt]} = D_t(U_{[t]}) - u_{tt} D_t(T) - u_{tx} D_t(X) - u_{ty} D_t(Y), \tag{3.136a}$$
$$U_{[tx]} = D_x(U_{[t]}) - u_{tt} D_x(T) - u_{tx} D_x(X) - u_{ty} D_x(Y), \tag{3.136b}$$
$$U_{[ty]} = D_y(U_{[t]}) - u_{tt} D_y(T) - u_{tx} D_y(X) - u_{ty} D_y(Y), \tag{3.136c}$$
$$U_{[xx]} = D_x(U_{[x]}) - u_{tx} D_x(T) - u_{xx} D_x(X) - u_{xy} D_x(Y), \tag{3.136d}$$

$$U_{[xy]} = D_x(U_{[y]}) - u_{ty}D_x(T) - u_{xy}D_x(X) - u_{yy}D_x(Y), \quad (3.136e)$$

$$U_{[yy]} = D_y(U_{[y]}) - u_{ty}D_y(T) - u_{xy}D_y(X) - u_{yy}D_y(Y). \quad (3.136f)$$

The total differential operators D_t , D_x and D_y are given by

$$D_t = \frac{\partial}{\partial t} + u_t\frac{\partial}{\partial u} + u_{tt}\frac{\partial}{\partial u_t} + u_{tx}\frac{\partial}{\partial u_x} + u_{ty}\frac{\partial}{\partial u_y} + u_{ttt}\frac{\partial}{\partial u_{tt}}\cdots$$

$$D_x = \frac{\partial}{\partial x} + u_x\frac{\partial}{\partial u} + u_{tx}\frac{\partial}{\partial u_t} + u_{xx}\frac{\partial}{\partial u_x} + u_{xy}\frac{\partial}{\partial u_y} + u_{ttx}\frac{\partial}{\partial u_{tt}}\cdots$$

$$D_y = \frac{\partial}{\partial y} + u_y\frac{\partial}{\partial u} + u_{ty}\frac{\partial}{\partial u_t} + u_{xy}\frac{\partial}{\partial u_x} + u_{yy}\frac{\partial}{\partial u_y} + u_{tty}\frac{\partial}{\partial u_{tt}}\cdots .$$

The associated invariant surfce condition is

$$Tu_t + Xu_x + Yu_y = U. \quad (3.137)$$

At this point, we will consider a particular example.

EXAMPLE 3.21

Consider the nonlinear diffusion equation

$$u_t = \left(uu_x\right)_x + \left(uu_y\right)_y. \quad (3.138)$$

The invariance condition of (3.138) is

$$U_{[t]} = uU_{[xx]} + Uu_{xx} + uU_{[yy]} + Uu_{yy} + 2u_xU_{[x]} + 2u_yU_{[y]}. \quad (3.139)$$

Substitution of the extended transformations (3.135) and (3.136) into the invariance conditions (3.139) gives, subject to the original equation gives the following set of determining equations:

$$T_x = 0, \quad T_y = 0, \quad T_u = 0, \quad X_u = 0, \quad Y_u = 0, \quad (3.140a)$$

$$X_x - Y_y = 0, \quad X_y + Y_x = 0, \quad (3.140b)$$

$$U + \left(T_t - 2X_x\right)u = 0, \quad (3.140c)$$

$$u^2U_{uu} + uU_u - U = 0, \quad (3.140d)$$

$$U_t - uU_{xx} - uU_{yy} = 0, \quad (3.140e)$$

$$2U_x + X_t + 2uU_{xu} - uX_{xx} - uX_{yy} = 0, \quad (3.140f)$$

$$2U_y + Y_t + 2uU_{yu} - uY_{xx} - uY_{yy} = 0. \quad (3.140g)$$

From (3.140b), we see that X and Y satisfy

$$X_{xx} + X_{yy} = 0, \quad Y_{xx} + Y_{yy} = 0.$$

Isolating U from (3.140c) gives

$$U = \left(2X_x - T_t\right) u, \tag{3.141}$$

from which it follows from (3.140e), (3.140f) and (3.140g) that

$$X_t = 0, \quad Y_t = 0, \quad X_{xx} = 0, \quad X_{xy} = 0 \quad U_t = 0, \tag{3.142}$$

and from (3.141) and the last equation in (3.142) $T_{tt} = 0$. Thus, solving (3.140b) and (3.142) we are lead to the following:

$$\begin{aligned} T = c_1 t + c_2, \quad & X = c_3 x + c_4 y + c_5, \\ Y = -c_4 x + c_3 y + c_6, \quad & U = (2c_3 - c_1)u. \end{aligned} \tag{3.143}$$

We now consider three separate reductions of the nonlinear diffusion equation (3.138). ■

EXAMPLE 3.21a Reduction 1 $c_1 = 1, c_i = 0\ \forall\, i \neq 1$

In this case, the invariant surface condition (3.137) becomes

$$tu_t = -u$$

whose solution is

$$u = \frac{f(x, y)}{t}.$$

Substitution in the original equation (3.138) gives

$$\left(ff_x\right)_x + \left(ff_y\right)_y + f = 0,$$

noting that we have reduced the number of independent variables. ■

EXAMPLE 3.21b Reduction 2 $c_1 = 2c_3$. The remaining $c_i = 0$.

In this case, the invariant surface condition (3.137) becomes

$$2tu_t + xu_x + yu_y = 0,$$

whose solution is

$$u = f\left(\frac{x}{\sqrt{t}}, \frac{y}{\sqrt{t}}\right).$$

Substitution in the original equation (3.138) gives

$$(ff_r)_r + (ff_s)_s + \frac{1}{2}rf_r + \frac{1}{2}sf_s = 0,$$

where $r = x/\sqrt{t}$ and $s = y/\sqrt{t}$.

■

EXAMPLE 3.21c Reduction 3 $c_3 = 1, c_i = 0, \forall\, i \neq 3$

In this case, the invariant surface condition (3.137) becomes

$$-yu_x + xu_y = 0,$$

whose solution is

$$u = f\left(x^2 + y^2, t\right).$$

Substitution in the original equation (3.138) gives

$$f_t - 4ff_r - 4rff_{rr} - 4rf_r^2 = 0,$$

where $r = x^2 + y^2$.

■

EXAMPLE 3.22

Here we calculate the symmetries of the equation

$$u_{xx} + u_{yy} + \left(e^u\right)_{zz} = 0, \tag{3.144}$$

which appears in quantum gravity (see Drew*et al.* [20] and the references within). Lie's invariance condition for (3.144) is (noting the new variables x, y and z)

$$U_{[xx]} + U_{[yy]} + e^u\left(U_{[zz]} + 2u_z U_{[z]}\right) + e^u\left(u_{xx} + u_z^2\right)U = 0. \tag{3.145}$$

The first- and second-order extensions are given by

$$U_{[x]} = D_x(U) - u_x D_x(X) - u_y D_x(Y) - u_z D_x(Z),$$
$$U_{[y]} = D_y(U) - u_x D_y(X) - u_y D_y(Y) - u_z D_y(Z),$$

$$U_{[z]} = D_z(U) - u_x D_z(X) - u_y D_z(Y) - u_z D_z(Z),$$
$$U_{[xx]} = D_x(U_{[x]}) - u_{xx} D_x(X) - u_{xy} D_x(Y) - u_{xz} D_x(Z),$$
$$U_{[yy]} = D_y(U_{[y]}) - u_{xy} D_y(X) - u_{yy} D_y(Y) - u_{yz} D_y(Z),$$
$$U_{[zz]} = D_z(U_{[z]}) - u_{xz} D_z(X) - u_{yz} D_z(Y) - u_{zz} D_z(Z).$$

Expanding (3.145) and imposing the original system equation, that is, $u_{xx} = -u_{yy} - \mathrm{e}^u \left(u_{zz} + u_z^2\right)$ gives on isolating the coefficients of u_x, u_y, u_z and all second-order derivatives gives

$$X_z = 0, \quad X_u = 0, \quad Y_z = 0, \quad Y_u = 0, \tag{3.146a}$$
$$Z_x = 0, \quad Z_y = 0, \quad Z_u = 0, \quad U_{uu} = 0, \tag{3.146b}$$
$$X_x - Y_y = 0, \quad X_y + Y_x = 0, \tag{3.146c}$$
$$U + 2X_x - 2Z_z = 0, \tag{3.146d}$$
$$U_u + U + 2X_x - 2Z_z = 0, \tag{3.146e}$$
$$2U_{xu} - X_{xx} - X_{yy} = 0, \tag{3.146f}$$
$$2U_{yu} - Y_{xx} - Y_{yy} = 0, \tag{3.146g}$$
$$2U_{zu} + 2U_z - Z_{zz} = 0, \tag{3.146h}$$
$$U_{xx} + U_{yy} + \mathrm{e}^u U_{zz} = 0, \tag{3.146i}$$

Upon introducing a potential function $f(x, y)$ such that

$$X = f_y, \quad Y = f_x, \tag{3.147}$$

shows that (3.146c) becomes

$$f_{xx} + f_{yy} = 0. \tag{3.148}$$

From (3.146d), we obtain U directly giving

$$U = 2Z_z - 2f_{xy}. \tag{3.149}$$

Substitution of (3.147) and (3.149) into the remaining determining equations (3.146) noting (3.148) gives

$$Z'' = 0,$$

which easily integrates giving

$$Z = c_1 z + c_0.$$

Thus, the infinitesimals for (3.144) are

$$X = f_y, \quad Y = f_x, \quad Z = c_1 z + c_0, \quad U = 2c_1 - 2f_{xy}, \tag{3.150}$$

where f satisfies (3.148). ∎

EXAMPLE 3.22a Reduction 1

If we choose $f = xy$ and $c_1 = 1$ in (3.150), then the invariant surface condition is

$$xu_x + yu_y + zu_z = 0.$$

The solution of this is

$$u = f\left(\frac{x}{z}, \frac{y}{z}\right),$$

and substitution into the original equation (3.144) gives

$$F_{rr} + F_{ss} + e^F\left(r^2F_{rr} + 2rsF_{rs} + s^2F_{ss} + (rF_r + sF_s)^2 + 2rF_r + 2sF_s\right) = 0,$$

where $r = x/z$ and $s = y/z$. ∎

EXAMPLE 3.22b Reduction 2

If we choose $f = y^2 - x^2$ and $c_1 = 1$ in (3.150), then the invariant surface condition is

$$yu_x - xu_y + zu_z = 2.$$

The solution of this is

$$u = 2\tan^{-1}\frac{y}{x} + f\left(x^2 + y^2, ze^{-\tan^{-1}\frac{y}{x}}\right),$$

and substitution into the original equation (3.144) gives

$$4r^2F_{rr} + \left(re^F + s^2\right)F_{ss} + 4rF_r + re^F F_s^2 + sF_s = 0,$$

where $r = x^2 + y^2$ and $s = ze^{\tan^{-1}\frac{y}{x}}$. ∎

EXERCISES

1. Calculate the symmetries for the Lin–Tsien equation

$$2u_{tx} + u_x u_{xx} - u_{yy} = 0,$$

(Ames and Nucci [21])

2. Calculate the symmetries for the Zabolotskaya–Khokhlov equation

$$u_{tx} - uu_{xx} - u_x^2 - u_{yy} = 0.$$

3. Calculate the symmetries for the Kadomtsev–Petviashvilli equation

$$\left(u_t + uu_x + u_{xxx}\right)_x + u_{yy} = 0.$$

Use a particular symmetry to reduce the PDE to an ODE.

4*. Classify the symmetries for the nonlinear diffusion equation

$$u_t = \left(f(u)u_x\right)_x + \left(g(u)u_y\right)_y.$$

(Dorodnitsyn *et al.* [22])

5*. Classify the symmetries for the nonlinear wave equation

$$u_{tt} = \left(f(u)u_x\right)_x + \left(g(u)u_y\right)_y.$$

6*. Classify the symmetries for the nonlinear diffusion equation

$$u_t = \nabla \cdot (D(u)\nabla u) - K'(u)u_z.$$

(Edwards and Broadbridge [23])

7*. Classify the symmetries for the nonlinear diffusion equation

$$u_t = u_{xx} + u_{yy} + Q\left(u, u_x, u_y\right).$$

(Arrigo *et al.* [24])

CHAPTER 4

Nonclassical Symmetries and Compatibility

4.1 NONCLASSICAL SYMMETRIES

In Chapter 3, we constructed the symmetries of the Boussinesq equation

$$u_{tt} + uu_{xx} + u_x^2 + u_{xxxx} = 0 \tag{4.1}$$

and found that they were

$$T = 2c_1 t + c_2, \quad X = c_1 x + c_3, \quad U = -2c_1 u.$$

The invariant surface condition is

$$\left(2c_1 t + c_2\right) u_t + (c_1 x + c_3) u_x = -2c_1 u, \tag{4.2}$$

and if $c_1 \neq 0$, then c_2 and c_3 can be set to zero without loss of generality as these constants just represent translation in t and x. This gives

$$2tu_t + xu_x = -2u, \tag{4.3}$$

which has as its solution

$$u = \frac{1}{t} f\left(\frac{x}{\sqrt{t}}\right). \tag{4.4}$$

Symmetry Analysis of Differential Equations: An Introduction,
First Edition. Daniel J. Arrigo.

Substituting (4.4) into the original equation (4.1) gives rise to the ODE

$$f^{(4)} + \left(f + \frac{r^2}{4}\right) f'' + f'^{\,2} + \frac{7}{4} r f' + 2f = 0,$$

where $r = x/\sqrt{t}$.

We now consider the following first-order PDE

$$u_t + t u_x = -2t, \tag{4.5}$$

noting that this cannot be obtained from (4.2). The solution of (4.5) is

$$u = -t^2 + f\left(x - \frac{t^2}{2}\right) \tag{4.6}$$

and substitution of (4.6) into the Boussinesq equation (4.1) gives

$$f^{(4)} + ff'' + f'^{\,2} - f' - 2 = 0,$$

another ODE! We have seen throughout Chapter 3 that a symmetry of a PDE can be used to reduce a PDE (in two independent variables) to an ODE. Therefore, it is natural to ask whether there is a symmetry explanation to this reduction. Before trying to answer this question, let us return back to the definition of a symmetry. A symmetry is a transformation that leaves a differential equation invariant. So, let us construct the Lie transformation group corresponding to (4.3) and (4.5). Recall, to obtain these transformations associated with the infinitesimals T, X and U, we need to solve

$$\frac{d\bar{t}}{d\varepsilon} = T\left(\bar{t}, \bar{x}, \bar{u}\right), \quad \frac{d\bar{x}}{d\varepsilon} = X\left(\bar{t}, \bar{x}, \bar{u}\right), \quad \frac{d\bar{u}}{d\varepsilon} = U\left(\bar{t}, \bar{x}, \bar{u}\right),$$

subject to $\bar{t} = t$, $\bar{x} = x$ and $\bar{u} = u$ when $\varepsilon = 0$. In the case of (4.3), we have

$$\frac{d\bar{t}}{d\varepsilon} = 2\bar{t}, \quad \frac{d\bar{x}}{d\varepsilon} = \bar{x}, \quad \frac{d\bar{u}}{d\varepsilon} = -2\bar{u},$$

which is easily solved giving

$$\bar{t} = e^{2\varepsilon} t, \quad \bar{x} = e^{\varepsilon} x, \quad \bar{u} = e^{-2\varepsilon} u.$$

A quick calculation shows that the Boussinesq equation is left invariant under this transformation. In the case of (4.5), we need to solve

$$\frac{d\bar{t}}{d\varepsilon} = 1, \quad \frac{d\bar{x}}{d\varepsilon} = \bar{t}, \quad \frac{d\bar{u}}{d\varepsilon} = -2\bar{t},$$

again subject to $\bar{t} = t$, $\bar{x} = x$, and $\bar{u} = u$ when $\varepsilon = 0$. The solution is given by

$$\bar{t} = t + \varepsilon, \quad \bar{x} = x + t\varepsilon + \frac{1}{2}\varepsilon^2, \quad \bar{u} = u - 2t\varepsilon - \varepsilon^2. \tag{4.7}$$

Under the transformation (4.7), the original equation

$$\bar{u}_{\bar{t}\,\bar{t}} + \bar{u}\,\bar{u}_{\bar{x}\,\bar{x}} + \bar{u}_{\bar{x}}^2 + \bar{u}_{\bar{x}\,\bar{x}\,\bar{x}\,\bar{x}} = 0$$

becomes

$$u_{tt} + uu_{xx} + u_x^2 + u_{xxxx} - 2\left(u_{tx} + tu_{xx}\right)\varepsilon = 0,$$

which is clearly not the original equation (4.1). However, if we impose the invariant surface condition (4.5), or more specifically, a differential consequence, we do get our original equation!

We now ask, can we seek the invariant of a particular PDE where not only do we impose the original equation but also the invariant surface condition. This idea was first proposed by Bluman [12] (see also Bluman and Cole [25]) and has been termed the "nonclassical method." It is important to note that the transformations associated with this method are *not* in fact symmetries at all but we use the term *nonclassical symmetries* loosely to be consistent with the literature.

Before proceeding, it is important to consider the invariance of the invariant surface condition.

4.1.1 Invariance of the Invariant Surface Condition

We consider the invariant surface condition

$$T(t,x,u)u_t + X(t,x,u)u_x = U(t,x,u) \tag{4.8}$$

and seek invariance under the infinitesimal transformations

$$\bar{t} = t + \varepsilon T(t,x,u) + O(\varepsilon^2),$$

$$\begin{aligned}\bar{x} &= x + \varepsilon X(t, x, u) + O(\varepsilon^2), \\ \bar{u} &= u + \varepsilon U(t, x, u) + O(\varepsilon^2).\end{aligned} \tag{4.9}$$

As usual, if we denote (4.8) by Δ, then invariance is given by

$$\Gamma^{(1)}\Delta|_{\Delta=0} = 0,$$

where Γ is the infinitesimal operator associated with (4.9) and $\Gamma^{(1)}$ its first extension. Thus, invariance is given by

$$\begin{aligned}&\left(TT_t + XT_x + UT_u\right)u_t + TU_{[t]} \\ &\quad + \left(TX_t + XX_x + UX_u\right)u_x + XU_{[x]} = TU_t + XU_x + UU_u,\end{aligned}$$

and substituting $U_{[t]}$ and $U_{[x]}$ from Chapter 3 gives

$$\begin{aligned}&\left(TT_t + XT_x + UT_u\right)u_t + T\left(U_t + u_tU_u - u_t(T_t + u_tT_u)\right. \\ &\left.-u_x(X_t + u_tX_u)\right) + \left(TX_t + XX_x + UX_u\right)u_x \\ &+ X\left(U_x + u_xU_u - u_t(T_x + u_xT_u) - u_x(X_x + u_xX_u)\right) \\ &\quad = TU_t + XU_x + UU_u.\end{aligned}$$

After cancellation

$$\begin{aligned}&\left(\cancel{TT_t} + \cancel{XT_x} + UT_u\right)u_t + T\left(\cancel{U_t} + u_tU_u - \cancel{u_tT_t} - u_t^2T_u - \cancel{u_xX_t} - u_tu_xX_u\right) \\ &+ \left(\cancel{TX_t} + \cancel{XX_x} + UX_u\right)u_x + X\left(\cancel{U_x} + u_xU_u - \cancel{u_tT_x} - u_tu_xT_u - \cancel{u_xX_x} - u_x^2X_u\right) \\ &\quad = \cancel{TU_t} + \cancel{XU_x} + UU_u,\end{aligned}$$

and rearrangement gives

$$\left(U - Tu_t - Xu_x\right)u_tT_u + \left(U - Tu_t - Xu_x\right)u_xX_u + \left(Tu_t + Xu_x - U\right)U_u = 0,$$

which is identically satisfied by virtue of the invariant surface condition (4.8).

4.1.2 The Nonclassical Method

We now seek invariance of the system

$$T(t,x,u)u_t + X(t,x,u)u_x = U(t,x,u), \tag{4.10a}$$

$$F(t,x,u,u_t,u_x,\dots) = 0. \tag{4.10b}$$

If we denote each equation in (4.10) by Δ_1 and Δ_2, then invariance is given by

$$\Gamma^{(1)}\Delta_1|_{\Delta_1=0,\Delta_2=0} = 0, \tag{4.11a}$$

$$\Gamma^{(n)}\Delta_2|_{\Delta_1=0,\Delta_2=0} = 0, \tag{4.11b}$$

but already established in the previous section, the first condition (4.11a) is identically satisfied and thus, we only consider the second condition (4.11b). Before we proceed, we find that some simplification can be made. If $T \neq 0$, then we can set $T = 1$ without loss of generality. The reason is that by imposing the invariant surface condition, we have at our disposal that the fact that

$$(k\,\Gamma)^{(n)} = k\,\Gamma^{(n)}. \tag{4.12}$$

It is clear from (4.12) that is true for $n = 0$. Here we show that this is true for $n = 1$ and leave it as an exercise for the reader to prove this for $n > 1$.

$$\begin{aligned}
(k\,\Gamma)^{(1)} &= k\,\Gamma + (k\,U)_{[t]}\frac{\partial}{\partial u_t} + (k\,U)_{[x]}\frac{\partial}{\partial u_x} \\
&= k\,\Gamma + \left(D_t(k\,U) - u_t D_t(k\,T) - u_x D_t(k\,X)\right)\frac{\partial}{\partial u_t} \\
&\quad + \left(D_t(k\,U) - u_t D_t(k\,T) - u_x D_t(k\,X)\right)\frac{\partial}{\partial u_x} \\
&= k\,\Gamma + k\left(D_t(U) - u_t D_t(T) - u_x D_t(X)\right)\frac{\partial}{\partial u_t} \\
&\quad + \overbrace{\left(U - Tu_t - Xu_x\right)}^{0} D_t(k)\frac{\partial}{\partial u_t} \\
&\quad + k\left(D_t(U) - u_t D_t(T) - u_x D_t(X)\right)\frac{\partial}{\partial u_x}
\end{aligned}$$

$$+\left(U - Tu_t - Xu_x\right)D_x(k)\frac{\partial}{\partial u_x}$$

$$= k\Gamma + k\left(D_t(U) - u_t D_t(T) - u_x D_t(X)\right)\frac{\partial}{\partial u_t}$$

$$+ k\left(D_t(U) - u_t D_t(T) - u_x D_t(X)\right)\frac{\partial}{\partial u_x}$$

$$= k\Gamma^{(1)}.$$

Now, because of this result, we can choose k as we wish. Here we choose $k = 1/T$ and rename X and U such that $X/T \to X$ and $U/T \to U$.

EXAMPLE 4.1

We first consider the heat equation originally considered by Bluman [12]. Lie's invariance condition for $u_t = u_{xx}$ is $U_{[t]} = U_{[xx]}$ or, with $U_{[t]}$ and $U_{[xx]}$ defined previously

$$\begin{aligned}
&U_t + \left(U_u - T_t\right)u_t - X_t u_x - T_u u_t^2 - X_u u_t u_x \\
&- U_{xx} + T_{xx}u_t - \left(2U_{xu} - X_{xx}\right)u_x + 2T_{xu}u_t u_x \\
&- \left(U_{uu} - 2X_{xu}\right)u_x^2 + T_{uu}u_t u_x^2 + X_{uu}u_x^3 \\
&+ 2T_x u_{tx} - \left(U_u - 2X_x\right)u_{xx} + 2T_u u_x u_{tx} \\
&+ T_u u_t u_{xx} + 3X_u u_x u_{xx} = 0.
\end{aligned} \tag{4.13}$$

As we have the flexibility of setting $T = 1$, we do so giving (4.13) as

$$\begin{aligned}
&U_t + U_u u_t - X_t u_x - X_u u_t u_x - U_{xx} - \left(2U_{xu} - X_{xx}\right)u_x \\
&- \left(U_{uu} - 2X_{xu}\right)u_x^2 + X_{uu}u_x^3 - \left(U_u - 2X_x\right)u_{xx} \\
&+ 3X_u u_x u_{xx} = 0.
\end{aligned} \tag{4.14}$$

This is subject to both the original equation $u_t = u_{xx}$ and the invariant surface condition (4.8) (with $T = 1$). Thus, we use

$$u_t = U - Xu_x, \quad u_{xx} = u_t = U - Xu_x. \tag{4.15}$$

Substitution of (4.15) into (4.14) gives

$$U_t - U_{xx} + 2UX_x + \left(X_{xx} - 2U_{xu} - X_t + 2UX_u - 2XX_x\right) u_x + \left(2X_{xu} - U_{uu} - 2XX_u\right) u_x^2 + X_{uu}u_x^3 = 0. \quad (4.16)$$

As before, the coefficients of the various powers of u_x in (4.16) are set to zero giving the determining equations

$$X_{uu} = 0, \quad (4.17a)$$

$$2X_{xu} - U_{uu} - 2XX_u = 0, \quad (4.17b)$$

$$X_{xx} - 2U_{xu} - X_t + 2UX_u - 2XX_x = 0, \quad (4.17c)$$

$$U_t - U_{xx} + 2UX_x = 0. \quad (4.17d)$$

From (4.17a), we find

$$X = A(t,x)u + B(t,x), \quad (4.18)$$

where A and B are arbitrary functions of their arguments. Substituting (4.18) into (4.17b) and integrating gives

$$U = -\frac{1}{3}A^2u^3 + \left(A_x - AB\right) u^2 + Pu + Q, \quad (4.19)$$

where P and Q are further arbitrary functions of t and x. Substituting (4.18) and (4.19) into (4.17c) and isolating coefficients of u gives

$$-\frac{2}{3}A^3 = 0, \quad (4.20a)$$

$$4AA_x - 2A^2B = 0, \quad (4.20b)$$

$$2AP + 2BA_x + 2AB_x - A_t - 3A_{xx} = 0, \quad (4.20c)$$

$$B_{xx} - B_t - 2BB_x + 2AQ - 2P_x = 0, \quad (4.20d)$$

from which we see that $A = 0$. This then leaves the entire system (4.20) as a single equation

$$B_t + 2BB_x - B_{xx} + 2P_x = 0. \quad (4.21)$$

Substituting (4.18) and (4.19) (with $A = 0$) into (4.17d) and isolating coefficients of u gives

$$P_t + 2PB_x - P_{xx} = 0, \quad (4.22a)$$

$$Q_t + 2QB_x - Q_{xx} = 0. \quad (4.22b)$$

Thus, (4.21) and (4.22) constitute three determining equations for the infinitesimals X and U given by

$$X = B(t, x), \quad U = P(t, x)u + Q(t, x).$$

Close observation with these determining equations shows that they are nonlinear and coupled! In fact, they were first obtained in 1967 by Bluman [12] and despite the many attempts to solve them in general, they remained unsolved for 30 years. In 1999, Mansfield [26] was the first to give the general solution of these equations, and later in 2002, Arrigo and Hickling [27] were able to show that the equations could be written as a matrix Burgers equation and hence linearizable via a matrix Hopf–Cole transformation. However, an earlier reference to their solution is given by Fushchych *et al.* [28].

So, at first sight, it appears that the nonclassical method has severe limitations. However, as we will see, that this in fact is not the case. ■

EXAMPLE 4.2

Consider the nonlinear heat equation

$$u_t = u_{xx} - 2u^3. \tag{4.23}$$

From the results from Chapter 3, this particular equation only admits a scaling and translational symmetry, that is,

$$T = 2c_1 t + c_2, \quad X = c_1 x + c_3, \quad U = -c_1 u.$$

Lie's invariance condition gives

$$U_{[t]} - U_{[xx]} + 6u^2 U = 0. \tag{4.24}$$

Substituting the appropriate extended transformations $U_{[t]}$ and $U_{[xx]}$ into (4.24) and both equation (4.23) and the invariant surface condition (4.8) gives rise to the following determining equations:

$$X_{uu} = 0, \tag{4.25a}$$

$$U_{uu} - 2X_{xu} + 2XX_u = 0, \tag{4.25b}$$

$$X_{xx} - 2U_{xu} - X_t + 2UX_u - 2XX_x + 6u^3 X_u = 0, \tag{4.25c}$$

$$U_t - U_{xx} + 2UX_x - 2u^3 U_u + 4u^3 U_u + 4u^3 X_x + 6u^2 U = 0. \tag{4.25d}$$

From the first two equations of (4.25), we find that

$$X = Au + B, \quad u = -\frac{1}{3}A^3u^2 + \left(A_x - AB\right)u + Pu + Q, \tag{4.26}$$

where A, B, P, and Q are functions of t and x. Substituting (4.26) into (4.25c) and isolating coefficients with respect to u gives

$$-\frac{2}{3}A\,(A-3)\,(A+3) = 0, \tag{4.27a}$$

$$2A\left(2A_x - AB\right) = 0, \tag{4.27b}$$

$$-A_t - 3A_{xx} + 2BA_x + 2AB_x + 2AP = 0, \tag{4.27c}$$

$$-B_t + B_{xx} - 2p_x - 2BB_x + 2AQ = 0, \tag{4.27d}$$

From (4.27a), we have three cases: $A = 0$, $A = 3$, and $A = -3$. The first case leads to the classical symmetries and will not be considered here. We focus on the second case and leave the third case to the reader. From (4.27b), we conclude that $B = 0$, from (4.27c), we then conclude that $P = 0$ and from (4.27d), we conclude that $Q = 0$. With these choices, the last determining equation (4.25d) is automatically satisfied. Thus, the infinitesimals are

$$X = 3u, \quad U = -3u^3.$$

The associated invariant surface condition is

$$u_t + 3uu_x = -3u^3,$$

whose solution is

$$x + \frac{1}{u} = F\left(6t - \frac{1}{u^2}\right), \tag{4.28}$$

where F is an arbitrary function. Substitution of (4.28) into the original equation (4.23) gives the ODE

$$F''(r) - 2F'(r)^3 = 0,$$

where $r = 6t - 1/u^2$, which is easily solved giving

$$F(r) = c_1 \pm \sqrt{c_2 - r},$$

where c_1 and c_2 are constants of integration. Thus, we end with the solution

$$x + \frac{1}{u} = c_2 \pm \sqrt{c_1 - 6t + \frac{1}{u^2}}. \tag{4.29}$$

As it turns out, we can solve (4.29) explicitly for u giving

$$u = \frac{2\left(x - c_2\right)}{c_1 - 6t - \left(x - c_2\right)^2}. \tag{4.30}$$

We note that from the classical symmetry analysis of (4.23) we can obtain the solution ansätz

$$u = \frac{1}{\sqrt{t}} F\left(\frac{x}{\sqrt{t}}\right),$$

which reduces the original PDE (4.23) to

$$2F''(r) + rF'(r) + F(r) - 4F^3(r) = 0, \tag{4.31}$$

where $r = x/\sqrt{t}$. However, it would be very difficult to obtain from (4.31) the solution

$$F(r) = \frac{2r}{r^2 + 6}$$

which corresponds to the solution (4.30) that we obtained from the nonclassical method. ■

EXAMPLE 4.3

Consider the Burgers' system

$$u_t = u_{xx} + uu_x - \frac{u^2}{v} v_x, \tag{4.32a}$$

$$v_t = v_{xx} + vv_x - \frac{v^2}{u} u_x. \tag{4.32b}$$

Applying the nonclassical method to the Burgers' system (4.32) we have the following determining equations:

$$X_{uu} = X_{vu} = X_{vv} = 0, \tag{4.33a}$$

$$2X_{xu} - 2uX_u + \frac{v^2}{u} X_v - U_{uu} - 2X_u X = 0, \tag{4.33b}$$

$$+2\frac{u^2}{v} X_u - vX_v - uX_v + 2X_{xv} - 2U_{vu} - 2X_v X = 0, \tag{4.33c}$$

$$-\frac{u^2}{v} X_v + U_{vv} = 0, \tag{4.33d}$$

$$-\frac{v^2}{u} X_u + V_{uu} = 0, \tag{4.33e}$$

$$2X_{xu} - uX_u - vX_u - 2V_{vu} + 2\frac{v^2}{u}X_v - 2X_uX = 0, \tag{4.33f}$$

$$2X_{xv} - 2X_vX + \frac{u^2}{v}X_u - 2vX_v - V_{vv} = 0, \tag{4.33g}$$

$$U_t + \frac{u^2}{v}V_x + 2UX_x - U_{xx} - uU_x = 0, \tag{4.33h}$$

$$V_t + \frac{v^2}{u}U_x - vV_x + 2VX_x - V_{xx} = 0, \tag{4.33i}$$

$$-U + X_{xx} - uX_x + 2X_uU - \frac{v^2}{u}U_v + \frac{u^2}{v}V_u$$
$$-2U_{xu} - X_t - 2XX_x = 0, \tag{4.33j}$$

$$-V + X_{xx} - vX_x + 2X_vV - \frac{u^2}{v}V_u + \frac{v^2}{u}U_v$$
$$-2V_{xv} - X_t - 2XX_x = 0. \tag{4.33k}$$

$$-2U_{xv} + 2UX_v + (v-u)U_v + \frac{u^2}{v}\left(X_x - U_u + V_v\right) + \frac{u}{v^2}$$
$$(2vU - uV) = 0, \tag{4.33l}$$

$$-2V_{xu} + 2VX_v + (u-v)V_u + \frac{v^2}{u}\left(X_x + U_u - V_v\right) + \frac{v}{u^2}$$
$$(2uV - vU) = 0. \tag{4.33m}$$

From (4.33a), we find that

$$X = A(t,x)u + B(t,x)v + C(t,x), \tag{4.34}$$

where A, B, and C are arbitrary smooth functions of t and x. Eliminating U from (4.33b)–(4.33c) using (4.34) gives

$$\frac{4v}{u}B - \frac{4u}{v}A - 2AB + B = 0$$

which must be satisfied for all u and v. Thus, $A = B = 0$. From (4.33b) and (4.33g), we see that

$$U_{uu} = U_{uv} = U_{vv} = V_{uu} = V_{uv} = V_{vv} = 0$$

giving that U and V are linear in u and v. Therefore,

$$U = P_1(t,x)u + P_2(t,x)v + P_3(t,x), \tag{4.35a}$$

$$V = Q_1(t,x)u + Q_2(t,x)v + Q_3(t,x), \tag{4.35b}$$

where P_i and Q_i, $i = 1, 2, 3$ are arbitrary smooth functions of t and x. With these, (4.35) and (4.33l) and (4.33m) become

$$-2P_{2x} + P_2 u + P_2 v + 2P_3 \frac{u}{v} + \left(P_1 + C_x\right) \frac{u^2}{v} - Q_3 \frac{u^2}{v^2} - Q_1 \frac{u^3}{v^2} = 0, \tag{4.36a}$$

$$-2Q_{1x} + Q_1 u + Q_1 v + 2Q_3 \frac{v}{u} + \left(Q_2 + C_x\right) \frac{v^2}{u} - P_3 \frac{v^2}{u^2} - P_2 \frac{v^3}{u^2} = 0 \tag{4.36b}$$

and since (4.36) must also be satisfied for all u and v gives

$$P_1 = Q_2 = -C_x, \quad P_2 = P_3 = Q_1 = Q_3 = 0.$$

Thus, the entire system (4.33) reduces to the following

$$C_t + 2CC_x - 3C_{xx} = 0, \tag{4.37a}$$

$$C_{tx} + 2C_x^2 - C_{xxx} = 0. \tag{4.37b}$$

Here, we find solutions to the overdetermined system (4.37). Eliminating the t derivative between (4.37a) and (4.37b) gives

$$C_{xxx} - CC_{xx} = 0. \tag{4.38}$$

Further requiring the compatibility of (4.37a) and (4.38) by calculating $\left(C_t\right)_{xxx}$ and $\left(C_{xxx}\right)_t$ gives

$$C_{xxxxx} - CC_{txx} - 4C_x C_{xxx} - 4C_{xx}^2 - C_t C_{xx} = 0, \tag{4.39}$$

and using (4.37a) and (4.38) to eliminate C_t and C_{xxx} and their differential consequences in (4.39) gives

$$C_{xx}\left(3C_{xx} - 2CC_x\right) = 0.$$

This gives rise to two cases: $C_{xx} = 0$ or $3C_{xx} - 2CC_x = 0$.

Case (1) $C_{xx} = 0$

In this case we solve $C_{xx} = 0$ giving $C = a(t)x + b(t)$ where a and b are arbitrary functions of integration. Substituting into (4.37a) and isolating coefficients with respect to x gives

$$\dot{a} + 2a^2 = 0, \quad \dot{b} + 2ab = 0,$$

each of which are easily solved giving

$$a = \frac{c_1}{2c_1 t + c_0}, \quad b = \frac{c_2}{2c_1 t + c_0},$$

where c_0, c_1, and c_2 are arbitrary constants. This leads to

$$C = \frac{c_1 x + c_2}{2c_1 t + c_0},$$

and via (4.34) and (4.35) we obtain

$$X = \frac{c_1 x + c_2}{2c_1 t + c_0}, \quad U = -\frac{c_1}{2c_1 t + c_0}, \quad V = -\frac{c_1}{2c_1 t + c_0}.$$

These would have risen from a classical symmetry analysis of our system (4.32) and hence are not true nonclassical symmetries.

Case (2) $3C_{xx} - 2CC_x = 0$

In this case, we see from (4.37a) that $C_t = 0$. Thus, we are required to find the common solution of

$$3C_{xx} - 2CC_x = 0, \quad C_{xxx} - 2C_x^2 = 0. \tag{4.40}$$

Eliminating C_{xxx} between the two equations in (4.40) and further simplifying gives

$$C_x \left(3C_x - C^2\right) = 0.$$

We omit the case where $C_x = 0$ as this would leave to a special case of Case (1) and thus focus on the second case $3C_x - C^2 = 0$ This is easily solve giving rise to

$$C = -\frac{3}{x + c},$$

where c is an arbitrary constant that we can set to zero without loss of generality. This, in turn, gives rise from (4.34) and (4.35) to the following nonclassical symmetries

$$X = -\frac{3}{x}, \quad U = -\frac{3u}{x^2}, \quad V = -\frac{3v}{x^2}. \tag{4.41}$$

Further, integrating the invariant surface conditions

$$u_t + Xu_x = U, \quad v_t + Xv_x = V,$$

with X, U, and V given in (4.41) gives

$$u = xP\left(x^2 + 6t\right), \quad v = xQ\left(x^2 + 6t\right), \tag{4.42}$$

and substitution of (4.42) into the original system (4.32) reduces it to

$$2QP'' + PQP' - P^2Q' = 0, \quad 2PQ'' + PQQ' - Q^2P' = 0. \tag{4.43}$$

Although no attempt will be made here to solve this system, any solution of (4.43) via (4.42) would give rise to a solution of (4.32). We note that the more general version of

$$u_t = u_{xx} + uu_x + F(u,v)v_x,$$
$$v_t = v_{xx} + vv_x + G(u,v)u_x,$$

was consider by Cherniha and Serov [29] and Arrigo *et al.* [7] and we refer the reader there for further details.

4.2 NONCLASSICAL SYMMETRY ANALYSIS AND COMPATIBILITY

For the nonclassical method, we seek invariance of both the original equations together with the invariant surface condition. For the previous example of the heat equation, if we let the invariant surface condition with $T = 1$ be Δ_1 and the heat equation be Δ_2, then

$$\Delta_1 = u_t + Xu_x - U, \tag{4.44a}$$

$$\Delta_2 = u_t - u_{xx}. \tag{4.44b}$$

We now reexamine the nonclassical method for the heat equation and show that the invariance condition

$$U_{[t]} = U_{[xx]} \tag{4.45}$$

arises naturally from a condition of compatibility. From (3.5) (with $T = 1$), we see that

$$U_{[t]} = D_tU - u_xD_tX = D_t\left(U - Xu_x\right) + Xu_{tx},$$
$$U_{[x]} = D_xU - u_xD_xX = D_x\left(U - Xu_x\right) + Xu_{xx},$$
$$U_{[xx]} = D_xU_{[x]} - u_{xx}D_xX = D_x^2\left(U - Xu_x\right) + Xu_{xxx},$$

so that (4.45) becomes

$$D_t\left(U - Xu_x\right) - D_x^2\left(U - Xu_x\right) + Xu_{tx} - Xu_{xxx} = 0,$$

which, by virtue of the heat equation, becomes

$$D_t\left(U - Xu_x\right) - D_x^2\left(U - Xu_x\right) = 0. \tag{4.46}$$

Rewriting the heat equation and the invariant surface condition equations, (4.44a) and (4.44b), as

$$u_{xx} = U - Xu_x, \tag{4.47a}$$

$$u_t = U - Xu_x, \tag{4.47b}$$

and imposing the compatibility condition

$$D_t\left(u_{xx}\right) - D_x^2\left(u_t\right) = 0,$$

gives rise to (4.46) naturally (see [30–33] for further details).

4.3 BEYOND SYMMETRIES ANALYSIS—GENERAL COMPATIBILITY

While both the classical and nonclassical symmetry methods have had tremendous success when applied to a wide variety of physically important nonlinear differential equations, there exist exact solutions to certain partial differential equations that cannot be explained using classical and nonclassical symmetry analyses. For example, Galaktionov [34] showed that the PDE

$$u_t = u_{xx} + u_x^2 + u^2 \tag{4.48}$$

admits the solution

$$u = a(t)\cos x + b(t), \tag{4.49}$$

where $a(t)$ and $b(t)$ satisfy the system of ODEs

$$\dot{a} = -a + 2ab, \qquad \dot{b} = a^2 + b^2. \tag{4.50}$$

A classical symmetry analysis of equation (4.48) leads to the determining equations

$$T_x = 0, \qquad T_u = 0, \qquad X_u = 0, \tag{4.51a}$$

$$2X_x - T_t = 0, \qquad U_{uu} + U_u = 0, \tag{4.51b}$$

$$U_t - U_{xx} + u^2 U_u - 2u^2 X_x - 2uU = 0, \tag{4.51c}$$

$$X_t - X_{xx} + 2U_{xu} + 2U_x = 0. \tag{4.51d}$$

A nonclassical symmetry analysis (with $T = 1$) of (4.48) leads to the system

$$X_{uu} - X_u = 0, \tag{4.52a}$$

$$U_{uu} + U_u - 2X_{xu} + 2XX_u = 0, \tag{4.52b}$$

$$2U_{xu} - 2UX_u + 3u^2 X_u + 2XX_u + X_t - X_{xx} + 2U_x = 0, \tag{4.52c}$$

$$U_t - U_{xx} + u^2 U_u + 2UX_x - 2u^2 X_x - 2uU = 0. \tag{4.52d}$$

Solving each system, (4.51) and (4.52), gives rise to

$$T = a, \qquad X = b, \qquad U = 0, \tag{4.53}$$

and

$$X = c, \qquad U = 0, \tag{4.54}$$

respectively, where a, b, and c are arbitrary constants. These are the same results as the associated invariant surface conditions for each are

$$au_t + bu_x = 0, \qquad u_t + cu_x = 0, \tag{4.55}$$

respectively, and scaling the first by a gives the second with $c = b/a$. However, this invariant surface condition will not give rise to the solution (4.49) illustrating that there are exact solutions of PDEs that cannot be obtained through classical and nonclassical symmetry analyses.

Despite the large number of success stories of both the classical and nonclassical symmetry methods, much effort has been spent trying to devise symmetry type methods to explain the construction of exact solutions to nonlinear PDEs such as the Galationov solution (4.49) for the diffusion equation (4.48). Olver [35] was the first to be able to obtain the Galationov solution by the method of differential constraints. He was able to show that by appending (4.48) with

$$u_{xx} - \cot x \, u_x = 0, \tag{4.56}$$

the Galationov solution could be obtained.

4.3.1 Compatibility with First-Order PDEs—Charpit's Method

The following is a derivation of Charpit's method. Consider the compatibility of the following first-order PDEs

$$F(x, y, u, p, q) = 0,$$
$$G(x, y, u, p, q) = 0,$$

where $p = u_x$ and $q = u_y$. Calculating second-order derivatives gives

$$\begin{aligned} F_x + pF_u + u_{xx}F_p + u_{xy}F_q &= 0, \\ F_y + qF_u + u_{xy}F_p + u_{yy}F_q &= 0, \\ G_x + pG_u + u_{xx}G_p + u_{xy}G_q &= 0, \\ G_y + qG_u + u_{xy}G_p + u_{yy}G_q &= 0. \end{aligned} \tag{4.57}$$

Solving the first three equations in (4.57) for u_{xx}, u_{xy} and u_{yy} gives

$$u_{xx} = \frac{-F_x\, G_q - p\, F_u\, G_q + F_q\, G_x + p\, F_q\, G_u}{F_p\, G_q - F_q\, G_p},$$

$$u_{xy} = \frac{-F_p\, G_x - p\, F_p\, G_u + F_x\, G_p + p\, F_u\, G_p}{F_p\, G_q - F_q\, G_p},$$

$$u_{yy} = \frac{\begin{array}{c} F_p^2\, G_x + p\, F_p^2\, G_u - F_y\, F_p\, G_q - q\, F_u\, F_p\, G_q \\ + q\, F_u\, F_q\, G_p - F_x\, F_p\, G_p - p\, F_u\, F_p\, G_p + F_y\, F_q\, G_p \end{array}}{F_p\, G_q - F_q\, G_p}.$$

Substitution into the last equation in (4.57) gives

$$F_p\, G_x + F_q\, G_y + (p\, F_p + q\, F_q)G_u - (F_x + p\, F_u)G_p - (F_y + q\, F_u)G_q = 0,$$

or conveniently written as

$$\begin{vmatrix} D_x F & F_p \\ D_x G & G_p \end{vmatrix} + \begin{vmatrix} D_y F & F_q \\ D_y G & G_q \end{vmatrix} = 0,$$

where $D_x\, F = F_x + p\, F_u$ and $D_y\, F = F_y + q\, F_u$. ■

EXAMPLE 4.4

Consider

$$u_t = u_x^2. \tag{4.58}$$

This is the example we considered already in Chapter 3 (Examples 3.1 and 3.2); however, now we will determine all classes of equations that are compatible with this one. Denoting

$$G = u_t - u_x^2 = p - q^2,$$

where $p = u_t$ and $q = u_x$, then

$$G_t = 0,\ \ G_x = 0,\ \ G_u = 0,\ \ G_p = 1,\ \ G_q = -2q,$$

and the Charpit equations are

$$\begin{vmatrix} D_tF & F_p \\ 0 & 1 \end{vmatrix} + \begin{vmatrix} D_xF & F_q \\ 0 & -2q \end{vmatrix} = 0,$$

or, after expansion

$$F_t - 2qF_x + \left(p - 2q^2\right) F_u = 0,$$

noting that the third term can be replaced by $-pF_u$ do to the original equation. Solving this linear PDE by the method of characteristics gives the solution as

$$F = F(x + 2tu_x,\ u + tu_t,\ u_t,\ u_x). \tag{4.59}$$

In Example 3.2, we found the invariant surface condition

$$tu_t + xu_x = u. \tag{4.60}$$

If we set F in (4.59) as

$$F(a, b, c, d) = ad - b,$$

and consider

$$F + 2t\left(u_t - u_x^2\right) = 0,$$

we obtain (4.60). From (4.59), we undoubtedly can an infinite number of compatible equations. ■

EXAMPLE 4.5

Consider

$$u_x^2 + u_y^2 = u^2. \tag{4.61}$$

Denoting $p = u_x$ and $q = u_y$, then

$$G = u_x^2 + u_y^2 - u^2 = p^2 + q^2 - u^2.$$

Thus,

$$G_x = 0,\quad G_y = 0,\quad G_u = 2u,\quad G_p = 2p,\quad G_q = 2q,$$

and the Charpit equations are

$$\begin{vmatrix} D_xF & F_p \\ -2pu & 2p \end{vmatrix} + \begin{vmatrix} D_yF & F_q \\ -2qu & 2q \end{vmatrix} = 0,$$

or, after expansion

$$pF_x + qF_y + \left(p^2 + q^2\right) F_u + puF_p + quF_q = 0, \tag{4.62}$$

noting that the third term can be replaced by u^2F_u do to the original equation. Solving (4.62), a linear PDE, by the method of characteristics gives the solution as

$$F = F\left(x - \frac{p}{u}\ln u, y - \frac{q}{u}\ln u, \frac{p}{u}, \frac{q}{u}\right).$$

Consider the following particular example

$$x - \frac{p}{u}\ln u + y - \frac{q}{u}\ln u = 0,$$

or

$$u_x + u_y = (x + y)\frac{u}{\ln u}.$$

If we let $u = \mathrm{e}^{\sqrt{v}}$, then this becomes

$$v_x + v_y = 2(x + y),$$

which has the solution

$$v = \frac{1}{2}(x + y)^2 + f(x - y).$$

This, in turn, gives the solution for u as

$$u = \mathrm{e}^{\sqrt{\frac{1}{2}(x+y)^2 + f(x-y)}}.$$

Substitution into the original equation (4.61) gives the following ODE

$$f'(r)^2 - 2f(r) = 0, \quad r = x - y.$$

Once this ODE is solved, then it can be used to construct an exact solution of the original PDE, (4.61).

It is interesting to note that when we substitute the solution of the compatible equation into the original it reduces to an ODE. A natural question is, does this always happen? This was addressed by Arrigo [36].

So just how complicated can things get when we try and generalize by seeking some fairly general compatibility. Our next example illustrates this. ■

EXAMPLE 4.6 General Compatibility of the Heat Equation

We seek compatibility of

$$u_t = u_{xx}, \tag{4.63a}$$

$$u_t = F\left(t, x, u, u_x\right), \tag{4.63b}$$

or

$$u_t = F\left(t, x, u, u_x\right), \quad u_{xx} = F\left(t, x, u, u_x\right), \tag{4.64}$$

where F is a function to be determined. Imposing compatibility $\left(u_t\right)_{xx} = \left(u_{xx}\right)_t$, expanding and imposing both equations in (4.64) gives

$$F_{xx} + 2pF_{xu} + 2FF_{xp} + p^2F_{uu} + 2pFF_{up} + F^2F_{pp} - F_t = 0, \tag{4.65}$$

where $p = u_x$. Clearly, (4.65) is complicated and nonlinear showing that compatibility might lead to a harder problem than the one we start with. However, any solution could be of possible use. For example, if we let $F = f(u)p^2$, then (4.65) becomes

$$f'' + 4ff' + 2f^3 = 0,$$

from which we see the general solution as

$$u = \int \frac{e^t}{\sqrt{c_1 t + c_2}}\,dt, \quad f = e^{-t}\sqrt{c_1 t + c_2}.$$

However, by choosing $c_1 = 0$ and $c_2 = 1$, we obtain explicitly

$$f = \frac{1}{u}.$$

Thus,

$$u_t = u_{xx}, \quad u_t = \frac{u_x^2}{u}$$

are compatible. ∎

EXAMPLE 4.7

In this example, we consider the compatibility between the $(2 + 1)$ dimensional reaction—diffusion equation (see Arrigo and Suazo [37])

$$u_t = u_{xx} + u_{yy} + Q(u, u_x, u_y), \tag{4.66}$$

and the first-order partial differential equation

$$u_t = F\left(t, x, y, u, u_x, u_y\right). \tag{4.67}$$

We will assume that

$$(F_{pp}, F_{pq}, F_{qq}) \neq (0, 0, 0), \tag{4.68}$$

where $p = u_x$ and $q = u_y$ as equality would give rise to the nonclassical method and we are trying to seek more general compatibility. We refer the reader to Arrigo *et al.* [24] for details on the classical and nonclassical symmetry analyses of (4.66).
Imposing compatibility $\left(u_t\right)_{xx} + \left(u_t\right)_{yy} = \left(u_{xx} + u_{yy}\right)_t$ between (4.66) and (4.67) gives rise to the compatibility equations

$$F_{pp} + F_{qq} = 0, \tag{4.69a}$$

$$F_{xp} - F_{yq} + pF_{up} - qF_{uq} + (F - Q)F_{pp} = 0, \tag{4.69b}$$

$$F_{xq} + F_{yp} + qF_{up} + pF_{uq} + (F - Q)F_{pq} = 0, \tag{4.69c}$$

$$\begin{aligned} &-F_t + F_{xx} + F_{yy} + 2pF_{xu} + 2qF_{yu} + 2(F - Q)F_{yq} \\ &\quad + \left(p^2 + q^2\right) F_{uu} + 2q(F - Q)F_{uq} + (F - Q)^2 F_{qq} \\ &\quad + Q_p F_x + Q_q F_y + \left(pQ_p + qQ_q - Q\right) F_u - pQ_u F_p - qQ_u F_q + FQ_u = 0. \end{aligned} \tag{4.69d}$$

Eliminating the x and y derivatives in (4.69b) and (4.69c) by (i) cross differentiation and (ii) imposing (4.69a) gives

$$2F_{up} + (F_p - Q_p)F_{pp} + (F_q - Q_q)F_{pq} = 0, \tag{4.70a}$$

$$2F_{uq} + (F_p - Q_p)F_{pq} + (F_q - Q_q)F_{qq} = 0. \tag{4.70b}$$

Further, eliminating F_{up} and F_{uq} by again (i) cross differentiation and (ii) imposing (4.69a) gives rise to

$$(2F_{pp} - Q_{pp} + Q_{qq})F_{pp} + 2(F_{pq} - Q_{pq})F_{pq} = 0, \tag{4.71a}$$

$$(Q_{pp} - Q_{qq})F_{pq} + 2Q_{pq}F_{qq} = 0. \tag{4.71b}$$

Solving (4.69a), (4.71a) and (4.71b) for F_{pp}, F_{pq}, and F_{qq}, respectively, gives rise to two cases:

$$\text{(i)} \quad F_{pp} = F_{pq} = F_{qq} = 0, \tag{4.72a}$$

$$\text{(ii)} \quad F_{pp} = \frac{1}{2}(Q_{pp} - Q_{qq}), \quad F_{pq} = Q_{pq}, \quad F_{qq} = \frac{1}{2}(Q_{qq} - Q_{pp}). \tag{4.72b}$$

As we are primarily interested in compatible equations that are more general than quasilinear, we omit the first case. It is interesting to note that if

$$Q_{pq} = 0, \quad Q_{pp} - Q_{qq} = 0, \tag{4.73}$$

then the second case becomes the first case. The solution of the overdetermined system (4.73) is

$$Q = Q_3(u)(p^2 + q^2) + Q_2(u)p + Q_1(u)q + Q_0(u), \tag{4.74}$$

for arbitrary functions $Q_0 - Q_3$. So, we will require that both equations in (4.73) are not satisfied.

If we require that the three equations in (4.72b) be compatible, we obtain

$$Q_{ppp} + Q_{pqq} = 0, \quad Q_{ppq} + Q_{qqq} = 0, \tag{4.75}$$

which integrates to give

$$Q_{pp} + Q_{qq} = 4a(u), \tag{4.76}$$

where a is an arbitrary function. However, under the transformation, $Q = \tilde{Q} + a(u)\left(p^2 + q^2\right)$ shows that $\tilde{Q}$ satisfies (4.76) with $a = 0$ and furthermore, to within equivalence transformation (a transformation of the original equation under $u = \phi(\tilde{u})$) shows that we can set $a = 0$ without loss of generality. Thus, Q satisfies

$$Q_{pp} + Q_{qq} = 0. \tag{4.77}$$

Using (4.77), we find that (4.72b) becomes

$$F_{pp} = Q_{pp}, \quad F_{pq} = Q_{pq}, \quad F_{qq} = Q_{qq}, \tag{4.78}$$

from which we find that

$$F = Q(u,p,q) + X(t,x,y,u)p + Y(t,x,y,u)q + U(t,x,y,u), \tag{4.79}$$

where X, Y, and U are arbitrary functions of their arguments. Substituting (4.79) into (4.70a) and (4.70b) gives

$$2Q_{up} + XQ_{pp} + YQ_{pq} + 2X_u = 0, \tag{4.80a}$$

$$2Q_{uq} + XQ_{pq} + YQ_{qq} + 2Y_u = 0, \tag{4.80b}$$

while (4.69b) and (4.69c) become (using (4.77) and (4.80))

$$(Xp + Yq + 2U)\,Q_{pp} + (Xq - Yp)\,Q_{pq} + 2\left(X_x - Y_y\right) = 0, \tag{4.81a}$$

$$(Xq - Yp)\,Q_{pp} - (Xp + Yq + 2U)\,Q_{pq} - 2\left(X_y + Y_x\right) = 0. \tag{4.81b}$$

If we differentiate (4.80a) and (4.80b) with respect to x and y, we obtain

$$X_xQ_{pp} + Y_xQ_{pq} + 2X_{xu} = 0, \quad X_xQ_{pq} + Y_xQ_{qq} + 2Y_{xu} = 0, \tag{4.82a}$$

$$X_yQ_{pp} + Y_yQ_{pq} + 2X_{yu} = 0, \quad X_yQ_{pq} + Y_yQ_{qq} + 2Y_{yu} = 0. \tag{4.82b}$$

If $X_x^2 + Y_x^2 \neq 0$, then solving (4.77) and (4.82a) for Q_{pp}, Q_{pq}, and Q_{qq} gives

$$Q_{pp} = -Q_{qq} = \frac{2\left(Y_xY_{xu} - X_xX_{xu}\right)}{X_x^2 + Y_x^2}, \quad Q_{pq} = -\frac{2\left(X_xY_{xu} + Y_xX_{xu}\right)}{X_x^2 + Y_x^2}.$$

If $X_y^2 + Y_y^2 \neq 0$, then solving (4.77) and (4.82b) for Q_{pp}, Q_{pq}, and Q_{qq} gives

$$Q_{pp} = -Q_{qq} = \frac{2\left(Y_yY_{yu} - X_yX_{yu}\right)}{X_y^2 + Y_y^2}, \quad Q_{pq} = -\frac{2\left(X_yY_{yu} + Y_yX_{yu}\right)}{X_y^2 + Y_y^2}.$$

In any case, this shows that Q_{pp}, Q_{pq}, and Q_{qq} are at most functions of u only. Thus, if we let

$$Q_{pp} = -Q_{qq} = 2g_1(u), \quad Q_{pq} = g_2(u),$$

for arbitrary functions g_1 and g_2, then Q has the form

$$Q = g_1(u)\left(p^2 - q^2\right) + g_2(u)p\,q + g_3(u)p + g_4(u)q + g_5(u), \tag{4.83}$$

where $g_3 - g_5$ are further arbitrary functions. Substituting (4.83) into (4.81) gives

$$2\left(Xp + Yq + 2U\right)g_1 + \left(Xq - Yp\right)g_2 + 2\left(X_x - Y_y\right) = 0, \tag{4.84a}$$

$$2\left(Xq - Yp\right)g_1 - \left(Xp + Yq + 2U\right)g_2 - 2\left(X_y + Y_x\right) = 0. \tag{4.84b}$$

As both equations in (4.84) must be satisfied for all p and q, this requires that each coefficient of p and q must vanish. This leads to

$$2g_1X - g_2Y = 0, \qquad g_2X + 2g_1Y = 0, \tag{4.85a}$$

$$2g_1U + X_x - Y_y = 0, \qquad g_2U + X_y + Y_x = 0. \tag{4.85b}$$

From (4.85a) we see that either $g_1 = g_2 = 0$ or $X = Y = 0$. If $g_1 = g_2 = 0$, then Q is quasilinear giving that F is quasilinear that violates our nonquasilinearity condition (4.68). If $X = Y = 0$, we are lead to a contradiction as we imposed $X_x^2 + Y_x^2 \neq 0$ or $X_y^2 + Y_y^2 \neq 0$. Thus, it follows that

$$X_x^2 + Y_x^2 = 0, \quad X_y^2 + Y_y^2 = 0,$$

or

$$X_x = 0, \quad X_y = 0, \quad Y_x = 0, \quad Y_y = 0.$$

As Q is not quasilinear then from (4.81) we deduce that

$$(Xp + Yq + 2U)^2 + (Xq - Yp)^2 = 0,$$

from which we obtain $X = Y = U = 0$. With this assignment, we see from (4.79) that $F = Q$ and from (4.80) that Q satisfies

$$Q_{up} = 0, \qquad Q_{uq} = 0, \tag{4.86}$$

which has the solution

$$Q = G(p,q) + H(u), \tag{4.87}$$

for arbitrary functions G and H. From (4.77), we find that G satisfies $G_{pp} + G_{qq} = 0$, whereas from (4.69d), we find that H satisfies $H'' = 0$ giving that $H = cu$ where c is an arbitrary constant noting that we have suppressed the second constant of integration due to translational freedom. This leads to our main result: equations of the form

$$u_t = u_{xx} + u_{yy} + cu + G\left(u_x, u_y\right),$$

are compatible with the first-order equations

$$u_t = cu + G\left(u_x, u_y\right).$$

■

4.3.2 Compatibility of Systems

Consider the nonlinear Schrodinger equation

$$i\psi_t + \psi_{xx} + \psi|\psi|^2 = 0. \tag{4.88}$$

If $\phi = u + iv$, then equating real and imaginary parts, we obtain the system

$$u_t + v_{xx} + v\left(u^2 + v^2\right) = 0, \tag{4.89a}$$

$$-v_t + u_{xx} + u\left(u^2 + v^2\right) = 0. \tag{4.89b}$$

For the nonclassical method, we would seek compatibility with (4.89) and the first-order equations

$$u_t + Xu_x = U, \quad v_t + Xv_x = V \tag{4.90}$$

where X, U, and V are functions of t, x, u, and v to be determined. However, one may ask whether more general compatibility equations may exist. For example, it is possible that equations of the form

$$u_t + Au_x + Bv_x = U, \quad v_t + Cu_x + Dv_x = V \tag{4.91}$$

(where A, B, C, D, U, and V are functions of t, x, u and v) exist such that (4.89) and (4.91) are compatible? Clearly, setting $A = D$ and $B = C = 0$ would lead to the nonclassical method but are there cases when $A \neq D$ and/or $B \neq C$. Trying to list here the determining equations and their subsequent analysis would be pointless as it would take pages and pages to complete. We however give the following two results. The first result is that if

$$A = D = \frac{c_1 x + 2c_2 t + c_3}{2c_1 t + c_0}, \quad B = C = 0$$

$$U = \frac{-c_1 u + \left(-c_2 x + c_5\right) v}{2c_1 t + c_0}, \quad V = \frac{\left(c_2 x - c_5\right) u - c_1 v}{2c_1 t + c_0},$$

where $c_0 - c_2$ are constant, then (4.91) becomes

$$\begin{aligned} u_t + \frac{c_1 x + 2c_2 t + c_3}{2c_1 t + c_0} u_x &= \frac{-c_1 u + \left(-c_2 x + c_5\right) v}{2c_1 t + c_0} \\ v_t + \frac{c_1 x + 2c_2 t + c_3}{2c_1 t + c_0} v_x &= \frac{\left(c_2 x - c_5\right) u - c_1 v}{2c_1 t + c_0} \end{aligned} \tag{4.92}$$

and (4.89) and (4.92) are compatible. This is nothing more than the nonclassical method. The second result is more interesting. If A, B, C, and D have the form

$$\begin{aligned} &A = c_0 v^2 + c_1 t + c_2, B = -c_0 uv, \\ &C = -c_0 uv, \quad D = c_0 u^2 + c_1 t + c_2, \\ &U = -\frac{1}{2} c_0 \left(c_1 t + c_2\right) v \left(u^2 + v^2\right) - \frac{1}{2}\left(c_1 x + c_3\right) v, \\ &V = \frac{1}{2} c_0 \left(c_1 t + c_2\right) u \left(u^2 + v^2\right) + \frac{1}{2}\left(c_1 x + c_3\right) u, \end{aligned}$$

then (4.91) becomes

$$\begin{aligned} u_t + \left(c_0 v^2 + c_1 t + c_2\right) u_x - c_0 uv v_x = &-\frac{1}{2} c_0 \left(c_1 t + c_2\right) v \left(u^2 + v^2\right) \\ &- \frac{1}{2}\left(c_1 x + c_3\right) v \end{aligned}$$

$$v_t - c_0 u v u_x + \left(c_0 u^2 + c_1 t + c_2\right) v_x = \frac{1}{2} c_0 \left(c_1 t + c_2\right) u \left(u^2 + v^2\right) + \frac{1}{2}\left(c_1 x + c_3\right) u$$

and these are compatible with (4.89). Although no attempt to solve these equations in general is presented, we note that by introducing polar coordinates $x = r\cos\theta$ and $y = r\sin\theta$, the two first-order compatible equations become

$$r_t + \left(c_1 t + c_2\right) r_x = 0$$

$$\theta_t + (c_0 r^2 + c_1 t + c_2)\theta_x = \tfrac{1}{2} c_0 (c_1 t + c_2) r^2 + \tfrac{1}{2}(c_1 x + c_3)$$

with the first one being linear.

4.3.3 Compatibility of the Nonlinear Heat Equation

We end this chapter with a question. Consider the nonlinear heat equation

$$u_t = \left(D(u) u_x\right)_x \tag{4.93}$$

or the system equivalent

$$v_t = D(u) u_x, \tag{4.94a}$$

$$v_x = u, \tag{4.94b}$$

then the nonclassical symmetries would be found by appending to this system the invariant surface conditions

$$u_t + X u_x = U, \tag{4.95a}$$

$$v_t + X v_x = V. \tag{4.95b}$$

If we let

$$U = XP + Q, \quad V = uX + DP, \tag{4.96}$$

then solving (4.94) and (4.95) for the first derivatives gives

$$u_t = Q, \quad u_x = P, \tag{4.97a}$$

$$v_t = DP, \quad v_x = u, \tag{4.97b}$$

and compatibility then gives rise to

$$P_t - Q_x + QP_u - PQ_u + DPP_v - uQ_v = 0, \tag{4.98a}$$

$$DP_x + DPP_u + uDP_v + D'P^2 - Q = 0. \tag{4.98b}$$

Would a symmetry analysis of this set of equation give rise to new classes of D? If we eliminate Q and obtain a second-order PDE, could again, new classes of D exist? These are interesting questions and require further study.

EXERCISES

1. Using the Charpit's method, find compatible equation with the following

$$\text{(i)} \quad u_x u_y = y$$

$$\text{(ii)} \quad u_x + u u_y^2 = 0$$

$$\text{(iii)} \quad u + u_x^2 + u_y = 0.$$

2. Calculate the nonclassical symmetries for the Burgers' equation

$$u_t + 2uu_x = u_{xx}. \tag{4.99}$$

3. Calculate the nonclassical symmetries for the Boussinesq equation

$$u_{tt} + uu_{xx} + u_x^2 + u_{xxxx} = 0 \tag{4.100}$$

(see Levi and Winternitz [38]).

4. Calculate the nonclassical symmetries for nonlinear heat equation

$$u_t = u_{xx} + \frac{2u^2}{x^2}(1-u) \tag{4.101}$$

(see Bradshaw-Hajek *et al.* [39]).

5. Calculate the nonclassical symmetries for boundary layer equation

$$u_y u_{xy} - u_x u_{yy} - u_{yyy} = 0 \tag{4.102}$$

(see Naz *et al.* [40]).

6. Calculate the nonclassical symmetries for the Burgers system

$$u_t = u_{xx} + uu_x + (u + a)v_x$$
$$v_t = v_{xx} + vv_x + (v + b)u_x,$$

where a and b are constants (see Arrigo *et al.* [7]).

7. Classify the nonclassical symmetries for the following

$$u_t = u_{xx} + Q(u)$$
$$u_t = \left(e^u u_x\right)_x + Q(u)$$
$$u_t = \left(u^n u_x\right)_x + Q(u)$$

(see Arrigo *et al.* [41], Arrigo and Hill [42] and Clarkson and Mansfield [43]).

8. Classify the nonclassical symmetries for the following

$$u_t = \nabla \cdot (D(u)\nabla u) + Q(u) \tag{4.103}$$

(see Goard and Broadbridge [44]).

4.4 CONCLUDING REMARKS

We are now at the end of our journey. This journey has introduced the reader to the method of symmetry analysis of differential equations and invariance. These symmetries have been shown to reduce their equations (ordinary and partial) to simpler ones. This chapter deals primarily with the nonclassical method and what's beyond—a light introduction to compatibility. So, are we really at the end of our journey or just the beginning? For the interested reader (which I hope), there are several places to go. First, as many may find, there are several more advanced books on the subject. We only mention a few. There are the books by Bluman and his collaborators: Bluman and Anco [45], Bluman *et al.* [46] and Bluman and Kumei [1]. The book by Cantwell, Cantwell [18] covering a lot of material from fluid mechanics, and the book by Olver [47]. Each book has its own strengths and should appeal to a lot of those wishing to learn more. I would also like to mention the three volume set, the *CRC Handbook of Lie Group Analysis of*

Differential Equations edited by Ibragimov [48–50]. It contains a vast amount of information in both materials and references.

So at this point, I bid you farewell and wish you well on the next part of your journey.

Solutions

Section 1.5

5. $a = 2b$

Section 2.2.1

1.(i) $X = y, \quad Y = x,$

1.(ii) $X = \dfrac{y}{x}, \quad Y = y,$

1.(iii) $X = -xy^2, \quad Y = -y^3.$

2.(i) $\bar{x} = x\mathrm{e}^{y\varepsilon + \frac{1}{2}\varepsilon^2}, \bar{y} = y + \varepsilon,$

2.(ii) $\bar{x} = x\mathrm{e}^{\varepsilon} + y\varepsilon\mathrm{e}^{\varepsilon}, \bar{y} = y\mathrm{e}^{\varepsilon},$

2.(iii) $\bar{x} = \dfrac{x}{x - (x-1)\mathrm{e}^{\varepsilon}}, \bar{y} = \dfrac{y}{-y + (y+1)\mathrm{e}^{-\varepsilon}}.$

3.(i) $X = c_1 x^3 + c_2 x, \quad Y = \left(c_1 x^2 + c_1 + c_2\right) x^2 y$

with $c_1 = 0, c_2 = 1, r = \dfrac{\mathrm{e}^{x^2}}{y^2}, \; s = \ln x, \; s' = -\dfrac{1}{2r(r+1)},$

3.(ii) $X = c_1 x, \quad Y = 3c_1 y$

with $c_1 = 1, r = x^3 y^{-1}, \; s = \ln x, \; s' = -\dfrac{r+2}{r^3},$

3.(iii) $X = c_1 y, \quad Y = 0$

with $c_1 = 1, r = y, \; s = \dfrac{x}{y}, s' = \dfrac{1}{r^2 + 1}.$

Symmetry Analysis of Differential Equations: An Introduction,
First Edition. Daniel J. Arrigo.

Section 2.3.6

1.(i) $X = 0, \quad Y = \dfrac{1}{x^3}, \quad x = r, \quad y = \dfrac{s}{r^3}, \quad \dfrac{ds}{dr} = e^r,$

1.(ii) $X = 0, \quad Y = e^{2x}, \quad x = r, \quad y = e^{2r}s, \quad \dfrac{ds}{dr} = re^{-2r},$

1.(iii) $X = 0, \quad Y = e^{-2x}y^2, \quad x = r, \quad y = -e^{2r}/s, \quad \dfrac{ds}{dr} = re^{2r},$

1.(iv) $X = 0, \quad Y = e^{-x^2}/y, \quad x = r, \quad y^2 = 2e^{-r^2}s, \quad \dfrac{ds}{dr} = r^3e^{r^2},$

1.(v) $X = x, \quad Y = y, \quad x = e^s, \quad y = re^s, \quad \dfrac{ds}{dr} = \dfrac{r}{r^3 - r^2 + 1},$

1.(vi) $X = x, \quad Y = y, \quad x = e^s, \quad y = re^s, \quad \dfrac{ds}{dr} = \dfrac{1}{\ln r - r},$

1.(vii) $X = -x, \quad Y = y, \quad \mu = \dfrac{1}{2xy},$

1.(viii) $X = 2x, \quad Y = y, \quad \mu = \dfrac{1}{7x^2y + 7xy^3},$

1.(ix) $X = 0, \quad Y = (y - x)^2, \quad x = r, \quad y = r - \dfrac{1}{s}, \quad \dfrac{ds}{dr} = \dfrac{1}{r^2},$

1.(x) $X = 0, \quad Y = (y - e^{-x})^2 e^{-3x}, \quad x = r,$

$$y = e^{-r} - \frac{e^{3r}}{s}, \quad \frac{ds}{dr} = e^{4r}.$$

Section 2.5.1

1. $a = -b, X = x, \quad Y = -y, \quad x = e^s, \quad y = re^{-s},$

$$s'' + \left(r^2 + (n-2)r\right)s'^3 + (3 - r - n)s'^2 = 0.$$

2. $X = \left(2c_1y + \dfrac{c_2}{y^2}\right)x^2 + \left(2c_3y^3 + \dfrac{c_4}{y^3} + c_5\right)x + c_6y^2 + \dfrac{c_7}{y}$

$$Y = \left(c_1y^2 - \frac{c_2}{y}\right)x + c_3y^4 + c_8y - \frac{c_4}{y^2}$$

from $c_2 = -1 \quad x = \left(\dfrac{r^2}{3s}\right)^{1/3}, y = (3rs)^{1/3} \quad rs'' + 2s' = 0$

from $c_5 = 1 \quad x = e^s, y = r \quad r^2 s'' + r^2 s'^2 - 2 = 0$

from $c_8 = 1 \quad x = r, y = e^s \quad s'' + 2rs'^3 + s'^2 = 0.$

3.(i) $n = -1, X = c_1 x^2 + c_2 x, \quad Y = c_1 xy, \quad x = \frac{1}{s}, \ y = \frac{r}{s},$

$rs'' + s'^3 = 0$

$n \neq -1, X = c_2 x, Y = 0, \quad x = e^s, \quad y = r, \quad s'' + r^n s'^3 + s'^2 = 0.$

3.(ii) $n = 1, X = c_1 x + c_2 x \ln x, \quad Y = -2(c_1 + c_2) - 2c_2 \ln x$

$x = e^{e^s}, \ y = r - 2e^s - 2s, s'' + (e^r - 2)s'^3 + s'^2 = 0$

$n \neq 1, X = c_1 x, Y = -2c_1, \quad x = e^s, \quad y = r - 2s,$

$s'' + (e^r + 2n - 2)s'^3 - (n-1)s'^2 = 0.$

4. $X = \left(c_1 y_1 + c_2 y_2\right) y + c_3 y_1^2 + c_4 y_2^2 + c_5,$

$Y = \left(c_1 y_1' + c_2 y_2'\right) y^2 + \left(c_3 y_1 y_1' + c_4 y_2 y_2' + c_6\right) y + c_7 y_1 + c_8 y_2$

where y_1 and y_2 are two independent solutions of $y'' + f(x)y = 0.$

Section 2.6.1

1.(i) $X = c_1 x + c_2, \quad Y = c_1 y, \quad$ if $c_1 = 1, c_2 = 0,$ then $x = e^s,$

$y = re^{-s}, s's''' - 3s''^2 + (r-6)s'^2 s'' + (6r - r^2)s'^5$

$+ (r - 11)s'^4 + s'^3 = 0.$

1.(ii) $X = c_2 x^2 + c_1 x + c_0, \quad Y = -\left(2c_2 x + c_1\right) y + 3c_2,$

if $X = x^2, Y = 3 - 2xy,$ then $x = \frac{1}{s}, \quad y = rs^2 + 3s,$

$s's''' - 3s''^2 - 4rs'^2 s'' + r^4 s'^5 - 6r^2 s'^4 + 3s'^3 = 0$

if $X = -x, Y = -y,$ then $x = e^s, \quad y = re^{-s},$

$s's''' - 3s''^2 + (4r - 6)s'^2 s'' - r(r-1)(r-2)(r-3)s'^5$

$-(6r^2 - 18r + 11)s'^4 - 3s'^3 = 0.$

1.(iii) $X = 4c_1 x + c_2, \quad Y = 3c_1 y,$ if $c_1 = 1, c_2 = 0,$ then $x = e^{4s},$

$y = r^{1/4} e^{3s}, -16r^2 s's''' + 48r^2 s''^2 + 48r^2 s'^2 s'' + 36rs's''$

$+64(15r + 64)r^2 s'^5 - 208r^2 s'^4 + 36rs'^3 + 21s'^2 = 0$

2. $X = c_2x^2 + c_1x + c_0, \quad Y = -(2c_2x + c_1)y - 6c_2.$
3. $X = c_1x + c_2, \quad Y = c_3y.$

Section 2.7.3

1. $T = -ct, \quad X = cx, \quad Y = cy, \quad Z = cz$
2. $T = -c_1t + c_2, \quad X = c_3y + c_4, \quad Y = -c_3x + c_5,$
3. $T = c_1, \quad X = c_2y, \quad Y = -c_2x,$
4. $T = c_1, \quad X = ac_2x - bc_2y + bc_3, \quad Y = ac_2y - bc_2x - ac_3,$
5. $T = c_1t + c_2, \quad X = -c_1x, \quad Y = -c_1y.$

where in 4, $c_1(a^2 - b^2) = 0$.

Section 3.1.4

(T, F, G are arbitrary)

1.(i) $T = T(t,x,u), \quad X = F(x-ct,u) + cT(t,x,u),$
$U = G(x-ct,u),$

1.(ii) $T = T(t,x,u), \quad X = F(x-ut,u) + tG(x-ut,u)$
$+ uT(t,x,u), \quad U = G(x-ut,u),$

1.(iii)
$$T = \left(c_1\,x^2 + c_2\,x + c_3\right) \mathrm{e}^t + c_6 + 2\,c_5\,x + 4\,c_4\,\mathrm{e}^{-t},$$
$$X = 2\,\left(2\,c_1\,x + c_2\right) u\mathrm{e}^t + \left(-4\,c_4\,x + 2\,c_7\right) \mathrm{e}^{-t} + 4\,c_5\,u - c_5\,x^2 + c_9\,x + c_{10}$$
$$U = \left(4\,c_1u^2 - \left(c_1\,x^2 + c_2\,x + c_3\right) u\right) \mathrm{e}^t - \left(4\,c_4\,u - c_4\,x^2 + c_7\,x - c_8\right) \mathrm{e}^{-t} - \left(4\,c_5\,x - 2\,c_9\right) u$$

1.(iv)
$$X = c_1u^2 + \left(2\,c_2\,x + c_3\right) u + 4c_4\,x^2 + c_5x + c_6,$$
$$Y = \left(c_1\frac{y}{x} + c_4x\right) u^2 + \left(c_3\frac{y}{x} + 4c_2y + c_7x\right) u + 4c_1\frac{y^2}{x} + c_6\frac{y}{x} + 4c_4xy + c_8x + 2c_9y$$

$$U = c_2u^2 + \left(\frac{4c_1y}{x} + 4c_4x + c_9\right)u + 2c_3\frac{y}{x} + 2c_7x$$
$$+ 4c_2y + c_{10}.$$

Section 3.2.5

1. $T = c_1 + c_2e^{2t} + c_3e^{-2t}$
   $X = \left(c_2e^{2t} - c_3e^{-2t}\right)x + c_4e^t + c_5e^{-t}$
   $U = -\left(c_2e^{2t}x^2 - c_3e^{-2t} + c_4e^tx + c_6\right)u + Q(t,x),\ Q_t = Q_{xx}$
   $+ (xQ)_x.$

2. $T = 4c_1t^2 + 2c_2t + c3$
   $X = (-c_1u^2 - c_4u - 2c_1t + c_5)x + Q(t,u),\ Q_t = Q_{uu}$
   $U = (4c_1t + c_2)u + 2c_4t + c_6.$

3.(i) $T = 2c_1t + c_2, \quad X = c_1x + c_3, \quad U = 0$
3.(ii) $T = 2c_1t + c_2, \quad X = c_1x - c_3u + c_4, \quad U = c_1u + c_3x + c_5.$

4. $T = c_1t + c_2, \quad X = c_3x + c_4, \quad U = 2(c_3 - c_1)u.$

5. $T = c_1, \quad X = c_2, \quad U = 0.$

6. $T = c_1, \quad X = c_2, \quad U = 0$

7. $X = c_1x + c_2y + c_3u + c_5$
   $Y = -c_2x + c_1y + c_4u + c_6$
   $U = -c_3x - c_4y + c_1u + c_7.$

Section 3.3.1

1.(i) $T = 3c_1t + c_2, \quad X = c_1x + c_3, \quad U = -c_1u$
1.(ii) $T = 3c_1t + c_2, \quad X = c_1x + 2c_3t + c_4, \quad U = -c_1u + c_3x + c_5$
1.(iii) $T = 3c_1t + c_2, \quad X = c_3x^2 + c_4x + c_5, \quad U = (2c_3x + c_4 - c_1)u.$

2. $X = c_1 x + c_2, \quad Y = (c_1 - c_3)y + f(x), \quad U = c_3 u + c_4.$

3. $T = c_1 t + c_2, \quad X = c_3 x + c_4, \quad U = (4c_3 - c_1)\frac{u}{n}.$

4. $T = c_1 t + c_2, \quad X = c_3 x + c_4, \quad U = (c_1 - 4c_3)\frac{u}{4}.$

Section 3.4.3

1. $T = 2c_1 t + c_2, \quad X = c_1 x + 2c_3 t + c_4,$
   $U = -c_1 u + (-c_3 x + c_5)v, \quad V = (c_3 x - c_5)u - c_1 v.$

2. $T = c_1 t^2 + 2c_2 t + c_3, \quad X = (c_1 t + c_2)x + c_4 t + c_5,$
   $U = \left(Fu + 2F_x\right) e^{v/4} - (c_1 t + c_2)u + c_1 x + c_4$
   $V = 4Fe^{v/4} + c_1 x^2 + 2c_4 x + 2c_1 t + c_6,$

where F satisfies $F_t = F_{xx}$.

Section 3.5.1

1. $T = 9a, \quad X = (3a' + 2c_1)x + 3a'' y^2 + 3b' y + c,$
   $Y = (6a' + c_1)y + 3b$
   $U = (-3a' + 4c_1)u + a^{(4)} y^4 + 2b''' y^3 + 6a''' xy^2 + 3a'' x^2 + 6b'' xy$
   $\quad + 2c'' y^2 + 2c' x + dy + e,$

where a, b, c, d, and e are arbitrary functions of t.

2. $T = 6a, \quad X = a'' y^2 + 2(a' + c_1)x + b' y + c, Y = (4a' + c_1)y + 2b$
   $U = 2(c_1 - 2a')u - a''' y^2 - 2a'' x - b'' y - c',$

where a, b, and b are arbitrary functions of t.

3. $T = 6a, \quad X = 2a' x - a'' y^2 - b' y + c, Y = 4a' y + 2b$
   $U = -4a' u - a''' y^2 + 2a'' x - b'' y + c',$

where a, b, and c are arbitrary functions of t.

Section 4.4

1. (i) $F\left(u_x, 2xu_x - u, u_y^2 - 2x, u_x u_y - y\right) = 0,$

(ii) $F\left(2x - \frac{1}{u_y^2}, \frac{u_x}{u_y}, uu_y, y - \frac{u}{u_y}\right) = 0,$

(iii) $F\left(2u_x + x, \ln u_y + y, \frac{u_x}{u_y}, u + u_x^2 + u_y\right) = 0.$

2. (i) $X = 2u, \quad U = 0,$

(ii) $X = \frac{(2c_2 t + c_1)x + 2c_3 t + c_4}{2c_2 t^2 + 2c_1 t + c_0}, \quad U = \frac{c_2 x + c_3 - (2c_2 t + c_1)u}{2c_2 t^2 + 2c_1 t + c_0},$

(iii) $X = -u + A(t, x), \quad Y = -u^3 + A(t, x)u^2 + B(t, x)u + C(t, x),$

where A, B, and C satisfy the following system

$$A_t + 2AA_x - A_{xx} + 2B_x = 0,$$
$$B_t + 2BA_x - B_{xx} + 2C_x = 0,$$
$$C_t + 2CA_x - C_{xx} = 0.$$

3. $X = ax + b,$

$U = -2au - 2a(\dot{a} + 2a^2)x^2 - 2(a\dot{b} + b\dot{a} + 4a^2 b)x - 2b(\dot{b} + 2ab),$

where $a = a(t)$ and $b = b(t)$ satisfy $\ddot{a} + 2a\dot{a} - 4a^3 = 0,\ \ddot{b} + 2a\dot{b} - 4a^2 b = 0$

4. (i) $X = \frac{c_1 x}{2c_1 t + c_0}, \quad U = 0$ (classical)

(ii) $X = \frac{3(u - 1)}{x}, \quad U = -\frac{3u(u - 1)^2}{x^2}.$

5. $Y = g'(x), \quad U = F(y - g(x)),$

where $F''' + FF'' - F'^2 = 0$. The rest are classical.

6. $X = \frac{u + v}{2} + A,$

$$U = -\frac{1}{4}(u+a)(u+v)^2 - \frac{b-a}{4}u^2 - \frac{A}{2}u(u+v) + \left(B - \frac{ab}{4}\right)$$
$$u - \frac{aA}{2}v + C,$$
$$V = -\frac{1}{4}(v+b)(u+v)^2 - \frac{a-b}{4}v^2 - \frac{A}{2}v(u+v) - \frac{bA}{2}u$$
$$+ \left(B - \frac{ab}{4}\right)v + D,$$

where A, B, C, and D satisfy the following system of equations:

$$A_t + 2AA_x + 2B_x - A_{xx} = 0,$$
$$B_t + 2BA_x - C_x - D_x - B_{xx} = 0,$$
$$C_t + 2CA_x - aD_x - C_{xx} = 0,$$
$$D_t + 2DA_x - bC_x - D_{xx} = 0.$$

References

1. Bluman GW and Kumei S, *Symmetries and Differential Equations*, Springer-Verlag, 1989.
2. Lie S, Zur Theorie des Integrebilit atsctor, *Christ. Forh.*, Aar **32** (1874) (Reprinted in Lie, S. *Gesammelte Abhandlungen*, Vol. 3 Paper XIII).
3. Bluman GW and Reid GJ, New symmetries of ordinary differential equations, *IMA J. Appl. Math.* **40** 87–94 (1988).
4. Clarkson PA and Olver PJ, Symmetry and the Chazy equation, *J. Diff. Eqns.* **124** (1) 225–246 (1996).
5. Bluman GW and Kumei S, On invariance properties of the wave equation, *J. Math. Phys.* **28** 307–318 (1987).
6. Sediawan WB and Megawati, Predicting chain length distribution in step growth polymerization by Monte Carlo method, *Am. J. Oil Chem. Tech.* **1** (9) 11–20 (2013).
7. Arrigo DJ, Ekrut DA, Fliss JR, and Le L, Nonclassical symmetries of a class of Burgers' system, *J. Math. Anal. Appl.* **371** 813–820 (2010).
8. Hopf E, The partial differential equation $u_t + uu_x = \mu u_{xx}$, *Commun. Pure Appl. Math.* **3** 201–230 (1950).
9. Cole JD, On a quasilinear parabolic equation occurring is aerodynamics, *Q. Appl. Math.* **9** 225–236 (1951).
10. Dorodnitsyn VA, On invariant solutions of equation of nonlinear heat conduction with a source, *USSR Comput. Math. Phys.* **22** 115–122 (1982).
11. Ovsjannikov LV, Gruppovye Svoystva Uravnenya Nelinaynoy Teploprovodnosty, *Dok. Akad. Nauk CCCP* **125** (3) 492 (1959).
12. Bluman GW, Construction of solutions to partial differential equations by the use of transformations groups, Ph.D. Thesis, California Institute of Technology (1967).
13. Edwards MP, Classical symmetry reductions of nonlinear diffusion-convection equations, *Phys. Lett. A* **190** 149–154 (1994).
14. Ames WF, Lohner RJ, and Adams E, Group properties of $u_{tt} = [f(u)u_x]_x$, *Int. J. Nonlinear Mech.* **16** 439–447 (1981).

Symmetry Analysis of Differential Equations: An Introduction,
First Edition. Daniel J. Arrigo.

15. Gandarias ML and Medina E, Analysis of a lubrication model through symmetry reductions, *Europhys. Lett.* **55** 143–149 (2001).

16. Broadbridge P and Tritscher P, An integrable fourth-order nonlinear evolution equation applied to thermal grooving of metal surfaces, *IMA J. Appl. Math.* **53** (3) 249–265 (1994).

17. Vinogradov AM and Krasil'shchik IS, On the theory of nonlocal symmetries of nonlinear partial differential equations, *Sov. Math. Dokl.* **29** 337–341 (1984).

18. Cantwell BJ, *Introduction to Symmetry Analysis, Cambridge Texts in Applied Mathematics*, Cambridge University Press, 2002.

19. Ma PKH and Hui WH, Similarity solutions of the two-dimensional unsteady boundary-layer equations, *J. Fluid Mech.* **216** 537–559 (1990).

20. Drew MS, Kloster SC, and Geigenberg JD, Lie group analysis and similarity solutions for the equation $u_{xx} + u_{yy} + (e^u)_{zz} = 0$, *Nonlinear Anal. Theor. Methods Appl.* **13** (5) 489–505 (1989).

21. Ames WF and Nucci MC, Analysis of fluid equations by group methods, *J. Eng. Mech.* **20** 181–187 (1985).

22. Dorodnitsyn VA, Knyazeva IV, and Svirshchevskii SR, Group properties of the heat equation with a source in two and three space dimensions, *Differ. Equ.* **19** 1215–1224 (1983).

23. Edwards MP and Broadbridge P, Exact transient solutions to nonlinear-diffusion-convection equations in higher dimensions, *J. Phys. A: Math. Gen.* **27** 5455–5465 (1994).

24. Arrigo DJ, Suazo LR, and Sule OM, Symmetry analysis of the two-dimensional diffusion equation with a source term, *J. Math. Anal. Appl.* **333** 52–67 (2007).

25. Bluman GW and Cole JD, The general similarity solution of the heat equation, *J. Math. Phys.* **18** 1025–1042 (1969).

26. Mansfield EL, Nonclassical group analysis of the heat equation, *J. Math. Anal. Appl.* **231** (2) 526–542 (1999).

27. Arrigo DJ and Hickling F, On the determining equations for the nonclassical reductions of the heat and Burgers' equation, *J. Math. Anal. Appl.* **270** 582–589 (2002).

28. Fushchych WI, Shtelen WM, Serov MI, and Popowych RO, *Q*-conditional symmetry of the linear heat equation, *Dokl. Akad. Nauk Ukrainy* **170** (12) 28–33 (1992).

29. Cherniha R and Serov M, Nonlinear systems of the Burgers-type equations: Lie and Q-conditional symmetries, Ansätze and solutions, *J. Math. Anal.* **282** 305–328 (2003).

30. Pucci E and Saccomandi G, On the weak symmetry groups of partial differential equations, *J. Math. Anal. Appl.* **163** 588–598 (1992).

31. Arrigo DJ and Beckham JR, Nonclassical symmetries of evolutionary partial differential equations and compatibility, *J. Math. Anal. Appl.* **289** 55–65 (2004).

32. Niu X and Pan Z, Nonclassical symmetries of a class of nonlinear partial differential equations with arbitrary order and compatibility, *J. Math. Anal. Appl.* **311** 479–488 (2005).
33. Niu X, Huang L, and Pan Z, The determining equations for the nonclassical method of the nonlinear differential equation(s) with arbitrary order can be obtained through the compatibility, *J. Math. Anal. Appl.* **320** 499–509 (2006).
34. Galaktionov VA, On new exact blow-up solutions for nonlinear equation with source and applications, *Differ. Integr. Equ.* **3** 863–874 (1990).
35. Olver PJ, Direct reduction and differential constraints, *Proc. R. Soc. London, Ser. A* **444** 509–523 (1994).
36. Arrigo DJ, Nonclassical contact symmetries and Charpit's method of compatibility, *J. Non. Math. Phys.* **12** (3) 321–329 (2005).
37. Arrigo DJ and Suazo LR, First Order Compatibility for (2+1)-dimensional diffusion equation, *J. Phys. A: Math Theor.* **41** 1–8 (2008).
38. Levi D and Winternitz P, Nonclassical symmetry reduction: example of the Boussinesq equation, *J. Phys. A: Math. Gen* **22** 2915–2924 (1989).
39. Bradshaw-Hajek BHY, Edwards MP, Broadbridge P, and Williams GH, Nonclassical symmetry solutions for reaction-diffusion equations with explicit spatial dependence, *Nonlinear Anal.* **67** 2541–2552 (2007).
40. Naz R, Khan MD, and Naeem I, Nonclassical symmetry analysis of the Boundary Layer Equations, *J. Appl. Math.* **2012** Article ID 938604, 7 (2012).
41. Arrigo DJ, Hill JM, and Broadbridge P, Nonclassical symmetry reductions of the linear diffusion equation with a nonlinear source, *I.M.A. J Appl. Math.* **52** 1–24 (1994).
42. Arrigo DJ and Hill JM, Nonclassical symmetries for nonlinear diffusion and absorption, *Stud. Appl. Math.* **94** 21–39 (1995).
43. Clarkson PA and Mansfield EL, Symmetry reductions and exact solutions of a class of nonlinear heat equation, *Physica D* **70** 250–288 (1994).
44. Goard J and Broadbridge P, Nonclassical symmetry analysis of nonlinear reaction-diffusion equations in two spatial dimensions, *Nonlinear Anal. Theor. Methods Appl.* **26** (4) 735–754 (1996).
45. Bluman GW and Anco SC, *Symmetry and Integration Methods for Differential Equations*, Springer-Verlag, 2002.
46. Bluman GW, Cheviakov AF, and Anco SC, *Applications of Symmetry Methods to Partial Differential Equations*, Springer-Verlag, 2010.
47. Olver PJ, *Applications of Lie Groups to Differential Equations*, 2nd Ed. Springer-Verlag, 1993.
48. Ibragamov NH, *Handbook of Lie Group Analysis of Differential Equations. Vol. 1: Symmetries, Exact Solutions and Conservation Laws*, CRC Press, Boca Raton, 1993.
49. Ibragamov NH, *Handbook of Lie Group Analysis of Differential Equations. Vol. 2: Applications in Engineering and Physical Sciences*, CRC Press, Boca Raton, 1994.

50. Ibragamov NH, *Handbook of Lie Group Analysis of Differential Equations. Vol. 3: New Trends in Theoretical Developments and Computational Methods*, CRC Press, Boca Raton, 1995.
51. Lie S, Klassifikation und Integration von gewohnlichen Differentialgleichen zwischen x, y die eine Gruppe von Transformationen gestatten, *Math. Ann.* **32** 213–281 (1888).

Index

Symmetry Analysis of Differential Equations: An Introduction,
First Edition. Daniel J. Arrigo.

Printed and bound by CPI Group (UK) Ltd, Croydon, CR0 4YY

27/10/2024

14580472-0005